T0291048

# Thinking Like a Physical Organic Chemist

# Thinking Like a Physical Organic Chemist

Steven M. Bachrach

OXFORD
UNIVERSITY PRESS

OXFORD
UNIVERSITY PRESS

Oxford University Press is a department of the University of Oxford. It furthers
the University's objective of excellence in research, scholarship, and education
by publishing worldwide. Oxford is a registered trade mark of Oxford University
Press in the UK and certain other countries.

Published in the United States of America by Oxford University Press
198 Madison Avenue, New York, NY 10016, United States of America.

© Oxford University Press 2023

CIP data is on file at the Library of Congress

ISBN 978–0–19–764037–1

DOI: 10.1093/oso/9780197640371.001.0001

Printed by Integrated Books International, United States of America

# Preface

In the late twentieth century and early twenty-first century, before Amazon took over the world, independent and chain booksellers had physical stores in malls and along downtown streets. They welcomed patrons to walk around and survey their titles. They encouraged you to sit and read for a while, offering comfy overstuffed chairs and a cappuccino from their in-house coffee bar.

I often enjoyed an hour's time browsing the aisles, hoping for serendipity to strike, offering me some unusual, good read. After checking out the new releases and fiction, I would make my way over to the science and technology sections. Even in those early days of the internet, the computing section was already running amok with books on programming in Java or HTML, introductions to online selling, or mastering programs like Excel and PowerPoint.

I would typically find three to four shelves' worth of physics titles. Authors and readers just can't seem to get enough of nontechnical interpretations of quantum mechanics, curved space, multiple dimensions, and time travel. There would always be a few biographies of Einstein, maybe a book about the Manhattan Project. Occasionally, I would come across Misner, Thorne, and Wheeler's monstrous *Gravitation*, a book I still plan/hope to read someday. This giant book weighs in at over six pounds and is 4 inches thick. It's actually quite amazing that any bookstore carried this behemoth at all.

The biology section would span a shelf or two. A reprint of Darwin's *The Origin of the Species* and some nontechnical approaches to evolution and genetics and maybe some fearmongering book proclaiming the threat of genetic engineering. Stephen Jay Gould would be well represented. Of course, there would be the creationists' alternative tracts. And there would be a handful of field guides to birds, trees, and reptiles.

Moving on to the chemistry section, invariably I was disappointed. At most, there was half a shelf worth of chemistry books. There would be a book to help you pass the chemistry portion of the MCAT (Medical College Admission Test) and maybe some forlorn introductory chemistry text. Where were the nonspecialist books describing the history and accomplishments of chemists? Why did physicists and biologists have all the fun? If chemistry is truly "the central science"—a subtitle of many an introductory textbook—how was it that no one had contributed a book explaining that role to the nonspecialist reader?

I understand that most people do not have fond memories of their interactions with chemistry in school. It is a difficult subject, heavily dependent on mathematics. Its vocabulary seems so foreign and impenetrable: monosodium glutamate, acetylsalicylic acid, butylated hydroxytoluene, polychlorinatedbiphenyl, dihydrogen monoxide. All of these sound dangerous and exotic, whether they are or not!

My own field of organic chemistry significantly contributes to the overall poor reputation of chemistry. Very few of our students are chemistry majors. Most students take a class in college-level organic chemistry because it is a requirement for their major, usually in preparation for a career in the health fields. Despite good intentions by the faculty, I'm sure that most students consider organic chemistry to be *the* premed weed-out class, designed solely to crash their dreams of attending medical school.

When I first meet a physician and they find out that I am a professor of organic chemistry, one of two responses follows. Some doctors rush to tell me how much they loved organic chemistry, how hard they worked, and how much they learned. The rest tell a different story, reflecting a disastrous experience. I recall a time that I visited a new dentist when my usual dentist was out sick. Right after I tell the dentist that I teach organic chemistry, he yells to his partner down the hall "Hey, I have an organic chemistry professor in my chair." The elicited response: "Give him a root canal for me!"

With this book I hope to help fill that gap on the bookstore shelf. I want you, the nonchemist reader, to appreciate what makes organic chemistry such a fascinating subject to those of us who have willingly spent our careers advancing this discipline. I will focus on how organic chemists think about what makes molecules react in the way they do. My emphasis will be on the nature of the logic within the construct of organic chemistry—how organic chemists have built a language and a grammar to understand and predict the behavior of organic molecules. Equally importantly, I'll disclose limitations in our language and models, and areas where uncertainty still pervades our thinking. Hopefully, you will recognize how the methods of organic chemists, their process of asking questions, devising tests, and building models, can translate to your area of interest.

It is this construction that explains why medical schools and veterinary schools require applicants to take a year of organic chemistry. Yes, organic chemistry provides the fundamental underpinnings of biological processes. However, most practicing physicians rarely, if ever, need to utilize organic chemistry as part of their diagnosis or treatment. What medical schools are really interested in is how a student performs in learning a new discipline from the ground up: a new vocabulary, new sets of rules, how to infer and extrapolate within that system. That experience is exactly what a student will face in medical school. Performance in organic chemistry may foretell how the student will take on the new language of medicine, the rules of how organs operate and interoperate, and how to predict outcomes from a series of symptoms and treatment options.

For the reader who may be a student in the first year organic chemistry sequence, this book provides an alternative way of thinking about organic chemistry. I am interested in providing a guide to *thinking* about organic chemistry, rather than teaching you how to solve problems in organic chemistry. Hopefully, this more philosophical approach may provide you a pathway toward developing your own chemical intuition.

The field of organic chemistry and its subdiscipline physical organic chemistry, which is what I will focus on most is too vast to tackle in one book. What I am providing is a flavor of what the field entails, a whetting of the appetite.

The language of organic chemistry rests on descriptions of the structure of molecules, how their constituent atoms are arranged and connected. I make use of these structure diagrams throughout the book with ample explanation so that the uninitiated reader can understand the point of each figure.

Mathematics plays an important role in our understanding of organic chemistry, as it does with all of the physical sciences. I only present a touch of mathematics, avoiding all use of calculus and higher mathematics. The intent again is to provide the science-inclined reader with a straightforward way in, primarily through analogy to real-world experiences.

I start the book with an introduction to reaction mechanisms, the central organizing concept for physical organic chemistry. Next, I present a few classic mechanisms, ones covered in every introductory organic chemistry class. Chapter 8 serves as an interlude, offering some thoughts on the scientific method in general. Then I revisit these earlier mechanisms and consider how we have simplified the models for the introductory courses, and then continue with a few more mechanisms. The penultimate chapter presents a new set of observations that are likely to lead to a revolution in the field, an overthrow of the mechanism paradigm prevalent today. The upshot is that science is ever evolving. That is the true joy of being a practicing scientist!

The challenges of the twenty-first century—climate change, pandemics, social justice—demand solutions that bring together expertise in a broad array of disciplines. Science and technology must play a collaborative role through which our partners broadly understand the scientific process, its limitations and values. Further, scientists must do better at communicating all aspects of what we do, so that society can properly leverage our contributions. The book concludes with a discussion of how lessons from the scientific method, as implemented by physical organic chemists, might provide some tools for tackling the challenges we face as a society. Do join me on this journey.

# Acknowledgments

This book emphasizes teaching and learning. So I begin the acknowledgments by thanking my own teachers who guided and influenced my education and ideas: Mr. Polster in 8th grade, my AP physics teacher, Mr. Hoepner, Gil Haight and Doug Applequist from the University of Illinois Chemistry Department, Harvey Stapleton from the University of Illinois Physics Department, and Clifford Dykstra, my undergraduate research mentor. From the University of California-Berkeley, I am indebted to Henry (Fritz) Schaefer, Clayton Heathcock, Peter Vollhardt, and Bob Bergman, all of whom helped me mature as a chemist. My fellow graduate students in the Streitweiser group provided hours of engaging discussions. My graduate research mentor, Andy Streitwieser, influenced me in so many positive ways, and his thinking pervades this book. Undoubtedly, my most important and most influential teacher was my father, Joseph Bachrach, an organic chemistry professor with an uncanny skill for explaining the science (and many, many other things).

I am grateful for conversations and lessons learned from many colleagues, including John Baldwin, Wes Borden, John Brauman, Bert Chandler, Chris Cramer, Jack Gilbert, Tom Gilbert, Jonathan Goodman, Stefan Grimme, Scott Gronert, Chris Hadad, Bill Hase, Ken Houk, Stephen Kass, Nancy Mills, Chris Purcell, Jim Ritchie, Henry Rzepa, Paul Schleyer, Peter Schreiner, Dan Singleton, Charlie Spangler, Bob Squires, Stu Staley, Dean Tantillo, Don Truhlar, and Adam Urbach.

I thank Monmouth University for resources to write this book, especially the assistance of the library staff. My Monmouth colleagues, Bill Schreiber and Koorleen Minton, provided excellent feedback on some of these chapters. I completed the final editing of the book at my new position at Radford University.

The Trivia Group of Jeanine and Ken Womack, Elyse and Jona Meer, and Mary and John Christopher helped raise all of our spirits during the Covid times. Their comments on a number of chapters were incredibly useful. Ken's insight and encouragement kept me committed to completing the book. It was Ken's constant enthusiasm that led me to Oxford University Press. I am especially grateful to Jeremy Lewis and his colleagues at OUP in seeing the value of my book.

The book benefited from excellent feedback from my son, Dustin. Many of the figures in the book were created by my talented sister, Lisa Bachrach. The person deserving the most thanks is my wife. Carmen insisted that since I was not taking on some new pandemic hobby or skill, like learning to paint or play the piano, I still had to make this time productive. Her insistence, and then unwavering support, truly made this book possible. While much of the Covid era has been painful, Carmen's love and support made it bearable and ultimately professionally productive.

# Contents

# 1
# Itineraries

Sue just received a promotion, but her new position requires her to move to London. She is very excited by the challenges of the new job. Sue always wanted to visit London, especially since she has never been out of the United States. Her whole life now seems open for new possibilities.

As she thinks about moving her possessions to England, she starts to ponder making this transition into an adventure. Why not take some time for a vacation in Europe on the way to London, she thinks. She can see in person so many of the cities she has only read about, visit some of the museums, and partake of the foods in their native lands. She starts putting together a list of some of her dream locales. Paris for sure, and Madrid for tapas and the Prado. Maybe Amsterdam to see a few Rembrandts and Van Goghs. How about Barcelona and Bordeaux? What about going to Germany, or Vienna, or Prague? And Rome, OMG! Her head starts spinning with options, and soon it becomes just so overwhelming. She reasonably can only take two weeks for the trip, and she doesn't want this to be some whirlwind experience, with just a short visit to each city. She needs to narrow down her list and make a plan!

Sue sits down and takes out a pen and a yellow legal pad. She writes down the names of all of the cities she's interested in seeing, one city per line. Her list just about fills the entire page. This is hopeless, she thinks. Ok, I'm going to start crossing names off this list! Slowly she draws a line through city name after city name. This is about the most depressing thing I've ever done, she says to herself. By dinnertime she has four names on her list and realizes that she gets London too! She decides to let it rest for the day and to finalize the list the next morning.

Sue awakes with an excitement to her early routine. I'm going to Europe! She sits down with her list and a coffee. This is it, she decides. I will visit Rome, Paris, Madrid, and Amsterdam. She takes out her laptop and logs into Expedia to figure out travel plans. Where to first, and then in what order should I visit each city? How many days in each city? How should I travel from one city to the next? She first decides that as a fan of *Mission Impossible*, she just has to take the Eurostar from Paris to London. That makes Paris her last stop. She looks at flights and decides that it's cheapest to fly to Rome from her home in

*Thinking Like a Physical Organic Chemist*. Steven M. Bachrach, Oxford University Press. © Oxford University Press 2023.
DOI: 10.1093/oso/9780197640371.003.0001

Brooklyn. Rome is so far away from these other cities, but she just has to get to Rome and see the Sistine Chapel and the Coliseum and all of those great places in *Roman Holiday*. Sue then realizes she can ride an overnight train from Rome to Madrid, so sleeping on the train won't mean missing any days of sightseeing. So that means she'll go from Madrid to Amsterdam, and the trip is set. Sue spends some time on how many days to spend in each city, finds hotels, and then books her air travel and trains. She writes down this final plan:

| | |
|---|---|
| Monday | Fly from JFK to Rome |
| Tuesday to Thursday | Visit Rome |
| Thursday night | Train from Rome to Madrid |
| Friday to Sunday | Visit Madrid |
| Sunday evening | Train from Madrid to Amsterdam |
| Monday to Wednesday | Visit Amsterdam |
| Wednesday afternoon | Train from Amsterdam to Paris |
| Wednesday to Saturday | Visit Paris |
| Sunday | Eurostar from Paris to London |

Sue now has a plan for her trip. A more precise term is *itinerary*, a schedule of where she is going, including the stops along the way. This itinerary can be further fleshed out by including airline flight numbers, train stations, departure and arrival times. She might add in the hotels for each night, as well as tickets for popular museums and reservations for dinners. More and more details can be added, accounting for more and more of how each day will be spent. Just how detailed, how granular an itinerary is a matter of choice and personality; should the plan account for everything or shall she leave some time for the serendipitous opportunity?

It is important to recognize that Sue's itinerary is not the only one that she could have developed. Remember that her only true constraints are where she is starting from (Brooklyn) and where she needs to end up (London, two weeks later). What cities she visits along the way, what order these cities are arranged in, are her choices.

As an organic chemist, Sue might look at this itinerary and see it as akin to a key element of her discipline. The itinerary is a list of points along a path from the starting location to the ending location. For the organic chemist, it's reformulated: how does this starting molecule transform into the final molecule? What molecules are produced along the way? Might there be other paths, other intermediate molecules, that could lead to that same final product? Just like an itinerary, more details could be added to flesh out this process and understand it in greater depth.

This notion of a step-by-step process for an initial molecule changing into a final molecule is called a *reaction mechanism*. This organizing principle allows an organic chemist to understand how molecules change over the course of a chemical reaction. More importantly, understanding a reaction mechanism allows us to predict how the process might alter under differing circumstances. Can we make the reaction run faster if we change the liquid (the solvent) we are using? Can we do something to obtain more product? What if we use a different starting molecule that is closely related to the original one—will it behave the same way?

The development of the concept of a reaction mechanism, along with the creation of instrumentation to gather data about the reaction, led to a complete transformation of the discipline of organic chemistry. The reaction mechanism created the conceptual framework for organic chemists to understand reactions, leading to new methods for transforming molecules and ultimately revolutionizing how molecules are synthesized. The paradigm of *reaction mechanism* changed the way organic chemists think, altering the language of the discipline. Organic chemistry became a mature discipline because of the *reaction mechanism*. The story told here highlights how the notion of reaction mechanism was created and highlights some of the seminal accomplishments along the way. It ends with the current stirrings among chemists that our notion of reaction mechanism may be too limited and that another revolution, another paradigm shift, may be in the making.

# 2

# Chemistry Basics

Chemistry is the study of the properties and transformations of substances such as water, salt, aspirin, plastic, sugar, steel, and rubber. Pretty much everything you interact with in your daily life are chemical substances. You get involved with chemistry frequently: when you cook food, when you bake bread, when the engine runs as you drive your car, when you get warmed by a fireplace or a gas furnace. These are all transformations of substances, what we call *chemical reactions*. The gasoline in your car's engine is ignited with oxygen to produce (mostly) carbon dioxide and water. The expansion of these product gases is what pushes the pistons and provides the energy to turn the wheels.

Organic chemistry is the study of organic compounds. Organic compounds are substances that contain the element carbon. Carbon is found in its pure form as graphite, the black solid in your pencil, or as diamond. It is much more commonly found in combinations with other elements, especially hydrogen, oxygen, and nitrogen. The name *organic chemistry* originates from the concept that compounds derived from living organisms (organic) were somehow special. However, in 1828 the German chemist Friedrich Wöhler prepared urea by combining silver cyanate with ammonium chloride. Urea is the compound in urine that gives urine its color and odor, and it is obviously the product of living organisms. But silver cyanate and ammonium chloride are inorganic compounds; we get them from refining nonliving materials dug from the ground. Wöhler's preparation of an organic substance without the intervention of some living process simply decimated the notion of a "vital force," some supernatural intervention that was necessary for the creation of life and its substances. So, even though we now routinely make "organic" molecules in test tubes and flasks, and not just in a living being, the name *organic chemistry* remains.

Organic chemistry is divided into two parts: synthetic organic chemistry and physical organic chemistry. *Synthetic organic chemistry* deals with the discovery of methods necessary to transform one compound into another one. It also involves multiple-step synthesis, such that when you start with common materials, you end up making complex, important molecules through what might be a long series of transformations.

*Thinking Like a Physical Organic Chemist.* Steven M. Bachrach, Oxford University Press. © Oxford University Press 2023. DOI: 10.1093/oso/9780197640371.003.0002

*Physical organic chemistry* is the study of why organic molecules behave the way they do. Why do some chemicals react very quickly, while seemingly small variations might retard the reaction? How do molecules come together and the elements change partners? Through application of a variety of observations—the rate of reaction, how much energy is produced, how much energy is needed for the reaction to occur, how much product is made—we hope to find insight into the way the reaction has proceeded. We are not interested solely in creating a theoretical framework for understanding reactions. There are at least two practical applications. First, through this theoretical framework, we may become better at predicting what will take place for a new reaction. Second, by understanding how a reaction takes place, we might be able to develop modifications that make the reaction faster, or yield more product, or increase the amount of one product over an alternative.

This is not a textbook. The intent is not to teach you organic chemistry. Rather, I hope that you will come to appreciate the logical structures that organic chemists have built in order to make sense of a very complicated and complex discipline. This logic structure demonstrates the beauty of human creativity in attempting to bring rationality and order to our view of Nature. The systematic logical processes developed over decades by physical organic chemists offer a model for how humans can address complex systems in disparate disciplines.

Rather than going through all of the gory physics, mathematics, and chemistry needed to understand fully organic chemical reaction mechanisms, I will present these topics largely through analogy to real-world everyday experiences and provide enough details that the reader can appreciate how this science has developed.

We will forego walking through the periodic table, the particles that make up an atom, the electronic structure of atoms, and all that orbital business. I will assume that you have some notion that all things in the universe are made up of molecules and molecules are made up of atoms. There are 118 (as of today) different types of atoms, called elements. For organic chemists, the most important element is carbon. All organic molecules have at least one carbon atom. So let's skip right to the heart of chemistry: the chemical bond.

The chemical bond keeps the neighboring atoms in a molecule together. The chemical bond is like a spring: stretch the ends of a spring and the spring pulls back, bringing the ends back to the original position. If you stretch a spring and let go, the ends will vibrate back and forth, and that's exactly what the two atoms of a bond are doing—they are constantly vibrating, approaching each other and then separating. The spring keeps the ends at nearly ideal separation: not too close, not too far apart, like Goldilocks preferring the bed that's not too soft and not too hard.

The chemical bond is best understood within the framework of quantum mechanics, which involves some heavy-duty mathematics. It is really quite beautiful math, and if one resides purely in that language of numbers and functions, the molecule and the chemical bond are well understood. The difficulty, as Physics Nobel Prize winner Werner Heisenberg noted in his book *Physics and Philosophy*, is that our language is constructed to make sense of and communicate the world in which we live, not that strange world of the subatomic where quantum mechanics rules the roost. The language we can use to describe the chemical bond must resort to models, analogies, and simplifications that often brush aside the intricacies of the mathematics. We are shoehorned into a position constrained by our classical world and its classical language.

How then to understand the chemical bond? What holds the atoms together in a molecule? What is that "chemical spring"? The atom is really mostly empty space, a sphere with almost all of the mass concentrated into a very small region at the center, called the nucleus. The nucleus is positively charged. Put two nuclei near each other and let go, and they will fly apart, driven by the repulsion between like charges. Electrical attraction is between a positively charged object and a negatively charged object. The negatively charged objects in atoms are the electrons, and these electrons fill up the vast majority of the volume of the atom, swirling about the nuclei.

The principles of quantum mechanics forbid us from knowing where the electron is or where it might be found some moments hence. Rather, we can only know the probability of finding the electron in a particular region. Instead of talking about position, we talk about electron density.

In a classical world, we might think that if we can put the electrons, with their negative charge, in between the positively charged nuclei, we might create a stable bound system. This situation is shown in Figure 2.1. In Figure 2.1a, the double-headed arrows represent the attraction between the oppositely

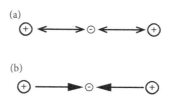

**Figure 2.1.** Model of electrostatic attraction in a bond: (a) the attraction for negative and positive charges and (b) the net force on each atom, which is positively charged.

charged objects, the positive nucleus for the negative electron. If we just focus on the forces acting on the nucleus, the left nucleus is pulled to the right by the attraction to the electrons, shown as the bold arrow to the right in Figure 2.1b. The nucleus on the right is pulled to the left by its attraction to the electrons; the bold arrow pointing leftward. These arrows point to one to the other nucleus, indicating that the nuclei are pulled together. For simplicity, I have neglected to draw the repulsive forces between the like charges of the nuclei. However, a stable system is one where the outward repulsion is balanced by the inward attraction.

The periodic table places elements in rows and columns based on their weights (more precisely by the number of protons in the nucleus) and chemical behaviors. The elements in a column share chemical reactivity. The elements of the farthest right column are most notable for their lack of reactivity. They are almost always found as solo atoms, rarely in molecules of any kind. This lack of reactivity suggests particular stability. Quantum mechanics provides an explanation for that stability related to the number of electrons in the atom. Of interest for organic chemists are the numbers of electrons in the two smallest inert elements, which are helium and neon. Helium has two electrons and neon has ten electrons. Any atom that can have two or ten electrons will attain some stability.

Now typically, we only look at the outer (also called *valence*) electrons, and that would be two for helium and eight for neon. So the very small atoms can have two electrons in two ways. For example, hydrogen can add one electron, resulting in a charge of −1, and lithium can lose an electron, creating a charge of +1. For atoms like carbon, nitrogen, and oxygen that appear in the same row of the periodic table as neon, they will want to have eight outer electrons. This could happen simply by, for example, oxygen picking up two electrons to add to its six, thereby resulting in a charge of −2. This is called filling the octet, or the *octet rule*.

For a carbon atom to fill its octet, it will need to gain four electrons, which would result in a charge of −4. That's a lot of charge to carry in such a small space. Might there be an alternative way to fill the octet? The chlorine molecule, composed of two chlorine atoms, $Cl_2$, provides some inspiration.

Every chlorine atom has seven outer electrons. Picking up one more electron would let the atom complete its octet. If we consider that process in the $Cl_2$ molecule, the left chlorine atom, say, might steal one electron from the right chlorine atom. The left atom would then have a complete octet, but the other chlorine atom would have only six valence electrons, a situation even worse than before! Perhaps each of the two chlorine atoms might provide one electron to share. This might be represented in Figure 2.2a, where each dot represents an electron.

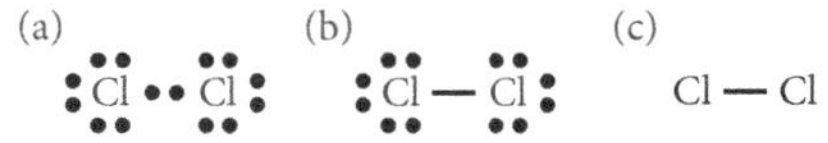

(a)                    (b)                  (c)

$$:\overset{..}{\underset{..}{Cl}}\cdot\cdot\overset{..}{\underset{..}{Cl}}:\qquad:\overset{..}{\underset{..}{Cl}}-\overset{..}{\underset{..}{Cl}}:\qquad Cl-Cl$$

**Figure 2.2.**  Representations of the bonding in the $Cl_2$ molecule.

This representation was developed by the great physical chemist G. N. Lewis, and it is called the Lewis dot structure. Each chlorine atom has eight electrons— six electrons that it owns outright and two from the shared pair in the middle.

Those two electrons positioned in between the two chlorine atoms represent the electrons that are shared. These shared electrons form the bond between the atoms. We call this a covalent bond, indicating a sharing of electrons. For simplicity, we will represent the bonding (shared) electron pair as a single line, as shown in Figure 2.2b. To simplify the representation further, we often omit the electrons that are unshared, as in Figure 2.2c, understanding that their presence is implied through completion of the octet.

The simplest organic compound contains just one carbon atom (re-member: all organic compounds possess at least one carbon) and four hy-drogen atoms. This is methane, $CH_4$, the combustible part of natural gas, which is often used for home cooking and heating. We can represent methane as

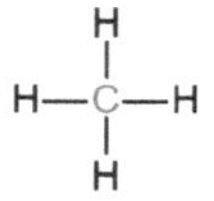

$$\begin{array}{c} H \\ | \\ H-C-H \\ | \\ H \end{array}$$

The carbon atom completes its octet by sharing two electrons with each hy-drogen, summing to eight. The hydrogen atoms are also satisfied, each having two electrons. The smallest organic molecule that has a carbon–carbon bond is ethane $C_2H_6$, represented as

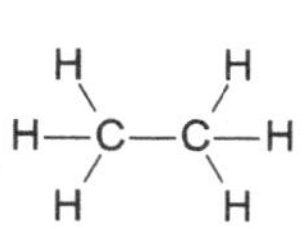

$$\begin{array}{cc} H \quad\quad H \\ \diagdown \quad\quad \diagup \\ H-C-C-H \\ \diagup \quad\quad \diagdown \\ H \quad\quad H \end{array}$$

Similar to methane, each carbon of ethane has a filled octet by making four bonds, each sharing two electrons between the bonded atoms.

It is possible for bonding atoms to share more than two electrons, forming multiple bonds. The ethene molecule $C_2H_4$ has a double bond between the two carbon atoms:

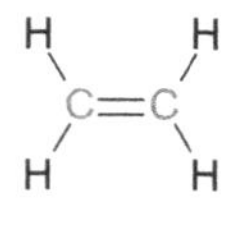

$$\begin{array}{cc} H \quad\quad H \\ \diagdown \quad\quad \diagup \\ C=C \\ \diagup \quad\quad \diagdown \\ H \quad\quad H \end{array}$$

Each carbon has a filled octet by sharing four electrons with each other and two electrons with two hydrogen atoms. Since there are four electrons in between the two carbon atoms, one might expect some physical consequences. More shared electrons might suggest a stronger bond. This is in fact the case, which results in the distance between the two atoms being shorter in a double bond than in a single bond. Organic compounds with a carbon–carbon double bond (often indicated as $C = C$) are categorized as *alkenes*.

It is possible for two carbons to be involved in a triple bond, with six electrons shared between the pair of atoms. The simplest example of this is ethyne, also called acetylene, $C_2H_2$, the combustible gas used in welding torches. Ethyne can be represented as

$$H-C\equiv C-H$$

Each carbon fills its octet by sharing six electrons with its neighboring carbon atom and two electrons with a hydrogen atom. As you might expect, the triple bond is stronger than the double and single bond, and it's shorter than these as well.

Chemistry is all about transforming a molecule into a different molecule. This is done in almost all cases by breaking bonds and making new bonds. We separate some atoms and then reattach them in some new way, transforming one type of molecule into another type, a different compound. The last century saw synthetic chemistry blossom with the development of some very powerful procedures for selectively breaking and making specific bonds.

Chemical reactions can be very simple or they can be quite complex. The simplest examples might be swapping one element for another, such as occurs in the reaction of chloromethane with hydrogen bromide to make bromomethane and hydrogen chloride:

The molecules on the left-hand side of the arrow are the reactants, and those on the right-hand side are the products. At first glance, it looks like we have just swapped the chlorine and bromine. In fact, as we'll see in Chapter 5, this is a *substitution reaction*—replacing the chlorine with bromine. When we look in terms of bonds, the following changes have occurred:

(a) Breaking the C–Cl bond
(b) Breaking the H–Br bond
(c) Making the C–Br bond
(d) Making the H–Cl bond

A chemical step defines which bonds are made and broken. This can be done one bond at a time, or multiple bond changes might occur collectively in one step.

Another example of a simple chemical reaction combines 1-butene and HBr to produce 2-bromobutane:

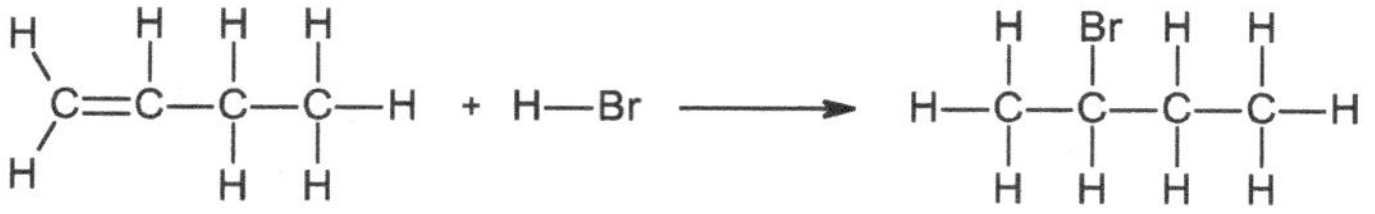

Let's again identify the bonds that have been broken and formed in this reaction:

(a) The double bond is broken, leaving the underlying single bond intact.
(b) The H–Br bond is broken.
(c) A hydrogen is attached to the terminal carbon.
(d) The C–Br bond is formed.

We call this type of reaction an *addition reaction* since two groups, H and Br, are added across the double bond.

We can also do the reverse of the above reaction but with a different reagent. Starting with 2-bromobutane and reacting it with the strong base sodium-tert-butoxide, we can make 1-butene, tert-butanol, and sodium bromide.

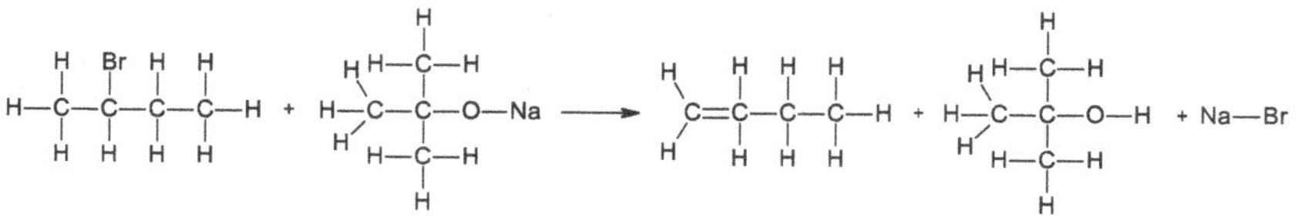

Three bonds are broken, C–H, C–Br, and O–Na, and three bonds are formed, Na–Br, O–H, and the second bond of the C–C double bond. This is an example of an *elimination reaction*, where two groups (in this case H and Br) are removed from adjacent carbon atoms to form a double bond.

I want to present a couple of notations that organic chemists use when writing reactions. Most reactions are run in solution, meaning as liquids. This approach is convenient for mixing and heating, and for insuring that all reactants can move around and find each other. The reactants and products are dissolved in the liquid, which is called the *solvent*. The solvent does not change, does not react, during the course of the reaction. It's just there to facilitate the reaction. We will often note the solvent by writing its name or structure above or more usually below the reaction arrow. Looking back at the substitution example above, we might write it as shown below, specifically indicating that acetonitrile $CH_3CN$ was used as the solvent.

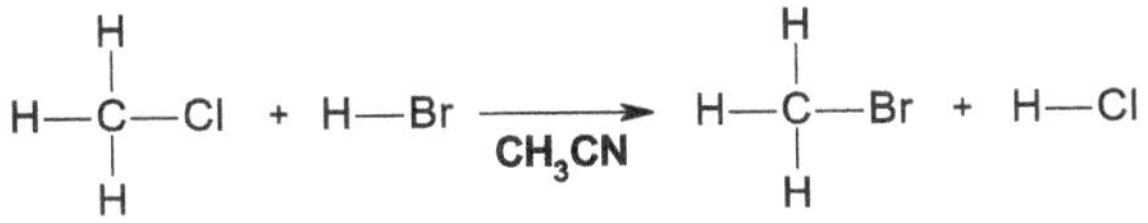

$$H-\overset{\displaystyle H}{\underset{\displaystyle H}{C}}-Cl \; + \; H-Br \xrightarrow[\textbf{CH}_3\textbf{CN}]{} H-\overset{\displaystyle H}{\underset{\displaystyle H}{C}}-Br \; + \; H-Cl$$

Often, some assistance is needed to speed up the reaction. This might be accomplished by heating it up or exposing the reaction flask to light. Sometimes another molecule might be added to the mixture. We call these added molecules *catalysts*. They help to increase the rate of reaction, but catalysts are overall unchanged during the reaction. They can be recovered after a reaction is completed, and they can be reused over and over again. Catalysts will be indicated above a reaction arrow, as in this last example.

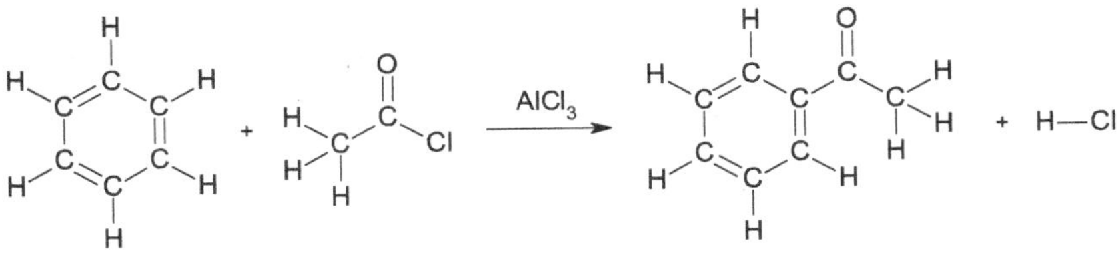

# 3
## *Tour de France* Stages

The *Tour de France* is the oldest and most famous of the grand tour bicycle races. Despite the recent spate of scandals associated with professional cycling, the *Tour de France* remains extremely popular, especially in France and the rest of Europe. The Tour takes place over twenty-three days, usually in July, in a series of twenty-one individual daily races called stages. Each stage may cover flat countryside, visit many quaint villages, pass by spectacular chalets and castles, or climb through the Pyrenees or Alps. The last stage ends with seven laps on the Champs-Élysées in Paris. The race has been run since 1903, except during both world wars. Not even the Covid-19 pandemic of 2020 could deter the race, though it was delayed until the end of August.

Stage 18 of the 2019 *Tour de France* took to the Alps. The race began in Embrun and finished some 208 km (129 mi) later in Valloire. The race was won by the Columbian cyclist Nairo Quintana, a specialist in mountain riding, in a time of 5 hours and 34 minutes.

The map of the stage course (Figure 3.1) shows the cities and some of the mountains crossed by the cyclists. (Note that Google estimates that this ride would take 15 hours and 10 minutes!) We can make a list of the towns that the cyclists visited on their way from Embrun to Valloire, an itinerary for this stage (Table 3.1). This is analogous to the "reaction mechanism," the step-by-step accounting for how the cyclists made their way through this part of France.

Although it may be daunting just to consider a 208 km (129 mi) bike ride, the truly amazing part of this stage was that it traversed three serious mountains: the Col de Vars, the famous Col d'Izoard, and finally the Col du Galibier, before a steep downhill ride into Valloire. The route over the Col de Vars crests at 2104 m (6903 ft), with a grade of 7.1%. The Col d'Izoard requires a 14 km (9 mi) climb with a grade of 7% over the top at 2354 m (7723 ft). The final climb over the Col du Galibier gets to 2622 m (8602 ft). Clearly, this day's race involved much more than just covering 208 km (129 mi) of pavement!

Shouldn't we have some way to represent both the distance traveled and all of this climbing so that we can understand just how grueling this stage was? Let's do this: let's stretch out the road traveled and make it into a straight line.

*Thinking Like a Physical Organic Chemist*. Steven M. Bachrach, Oxford University Press. © Oxford University Press 2023.
DOI: 10.1093/oso/9780197640371.003.0003

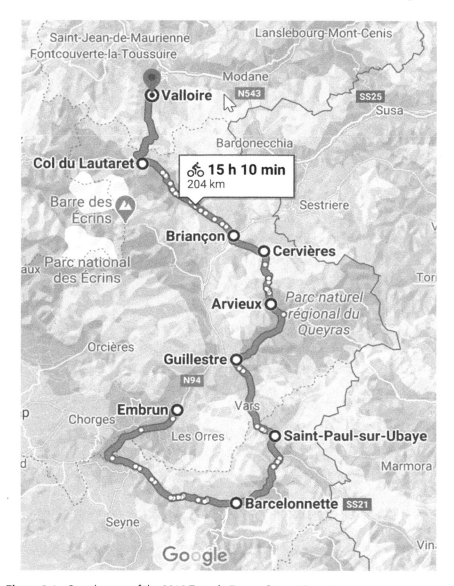

**Figure 3.1.** Google map of the 2019 Tour de France Stage 18 route.

Then we'll mark the towns along that line representing the distance traveled, creating Figure 3.2. This will become the *x*-axis of our new plot.

Next, we will plot on the *y*-axis the height above sea level for each city along the ride. We'll also include the elevations of the high points of the mountains. For the moment, we will just connect these elevations, giving us the plot in Figure 3.3. This plot clearly indicates some significant uphill pedaling

**Table 3.1**  The Itinerary of Stage 18

| Town | Distance (km) |
| --- | --- |
| Embrun | 0.0 |
| Savines-le-Lac | 6.5 |
| Le Lauzet-Ubaye | 31.5 |
| Barcelonnette | 51.0 |
| Jausiers | 60.5 |
| Saint-Paul-sur-Ubaye | 73.5 |
| Guillestre | 102.5 |
| Arvieux | 122.5 |
| Cervieres | 142.5 |
| Briancon | 152.0 |
| Serre-Chevalier | 161.0 |
| Minetier-les-Bains | 167.0 |
| Valloire | 208.0 |

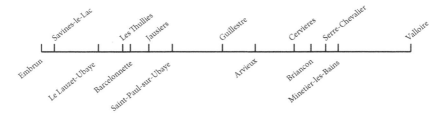

**Figure 3.2.**  Distances of cities along Stage 18 represented on a line.

that these cyclists had to do! If we had more data—the elevations for more locations along the route, for example—we could fill in the sections between the towns more accurately, thereby visualizing the steepness of each ascent and descent.

Traversing this stage is more than just riding a long way. The key to finishing this ride is having the energy necessary to climb over the mountains. You need to fight gravity all the way up each mountain; three times you have to climb above 2100 m (6890 ft). These three mountains were the day's challenge, and set the tone for the stage and its outcome.

How is the plot presented in Figure 3.3 relevant to organic chemistry? In the previous chapter, I discussed chemical steps in a reaction. It turns out that each chemical step requires energy to proceed. Just as the cyclist who has to

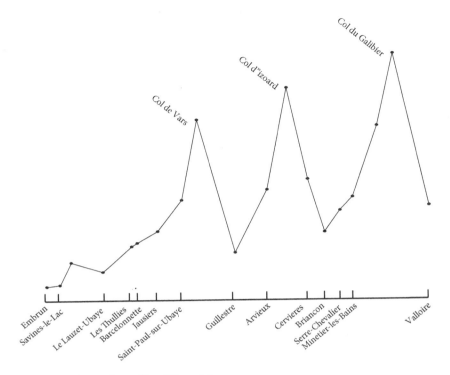

**Figure 3.3.** Schematic profile of Stage 18.

find the energy to climb over the peak, so the molecules need sufficient energy to get the reaction to occur.

Why is energy needed to make a reaction take place? Here we'll examine three different requirements of a typical chemical step, each of which has an associated energy demand. Together, these requirements will create an energy barrier that must be overcome for the reaction to proceed.

In Chapter 2, I stated that a chemical reaction involves the making and breaking of chemical bonds. The chemical bond is an energy storage device. The atoms in a bond are held to each other through electrostatic attraction between the positively charged nucleus in the center of the atom and the electrons distributed especially in the region between the atoms.

To pull the atoms apart, to break the bond, is like pulling on a rubber band. You have to supply energy to stretch the rubber band. Similarly, to break a chemical bond, energy must be supplied to pull the atoms apart from each other and snap the bond. The electrostatic attraction between the atoms must be overcome by pulling against the electrostatic force, trying to return the atoms to their original positions.

We can represent the energy of a bond with a Morse Potential, as shown in Figure 3.4. This representation was proposed by the American physicist Phillip Morse, best known as a scientific administrator—he was the first director of Brookhaven National Laboratory—but he was also an accomplished physicist. Along the $x$-axis is the distance between the two atoms of the bond. When the distance is too short, the electrons of the core regions of each atom get too close and repel each other. This is reflected by the dramatic rise in energy on the $y$-axis as the distance gets very short. As the bond is stretched, it is weakened by the increasing separation between the opposite charges of the electrons and the nuclei, and the energy increases. Bond breaking is moving the atoms apart, or moving along the $x$-axis to the right. In order to do that, the energy must increase, and that energy has to be supplied to make the reaction take place. This is seen in Figure 3.4 with the rising energy and then a flattening of the curve with increasing distance as the bond is completely broken.

In order to account for the other sources of the energy barrier associated with a chemical step, we need to make a short diversion to discuss entropy. When you first open up a brand-new deck of cards, the fifty-two cards are in perfect order, increasing from two through ten and then the picture cards ending in the ace, one suit at a time. Now shuffle the deck, and shuffle it a few more times. The cards are in a random order, with no noticeable pattern.

Imagine your child's room at the beginning of a day home from school. All the clothes are in their drawers, the desk is tidy, and the toys are all stored in their proper boxes stacked neatly on a shelf. This room looks very different at the end of the day. There are at least two sets of the day's clothes thrown on the floor. A few books, some paper, a set of colored pencils are scattered on the desk. On top of the dresser sits a plate and glass and the remains of the

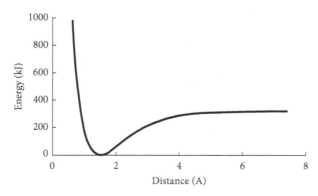

**Figure 3.4.** Morse Potential representation of a typical chemical bond.

afternoon snack. There are open boxes of games on the floor and toys strewn on the bed.

You are carrying very carefully your favorite glass vase, a present from an aunt from ages ago. In a blink of an eye, the vase slips from your grasp, falls to the floor, and smashes into what looks like a million pieces.

These three scenarios depict the ravages of entropy, a natural movement from order to disorder. Entropy is a measure of disorder or randomness. A just-opened deck of cards is highly ordered; the shuffled deck has increased entropy. The tidy room is nicely ordered; the messy room is disordered. The intact vase is highly ordered; the shattered vase has large entropy. It is a property of the universe we live in that entropy increases over time. Left to its own devices, disorder will arise. Think of the slow disappearance of a small city in the desert to ghost town to forlorn remnants of chimneys as a metaphor for where the universe is headed.

The American physicist Willard Gibbs recognized the connection between entropy and energy, immortalized in the equation that bears his name. To understand this connection, think about undoing the three scenarios I just presented.

You grab the shuffled deck of cards and start separating them into four separate piles, one for each suit. Then you take the thirteen clubs and put them in order, and repeat this manual process for the remaining three suits. For your child's room, you contemplate which solution is easier: supervising your child's efforts to clean their own room or just doing it yourself. You opt for the latter, and pick up the dirty laundry from the floor and place it into the hamper. You rearrange the top of her desk. You take the dirty dishes downstairs to the kitchen and put them in the dishwasher. Lastly, you box up all of the games and toys and return them to their spots on the shelves. Now to the broken vase; well, entropy has likely won that battle. Probably no amount of work can be done to reassemble the shattered vase.

The element common to reversing the disorder is that work has to be done, that real energy needs to be expended to sort the cards, clean the room, or reassemble the vase. Any activity that makes cleanliness from mess, that replaces randomness with predictability, that makes order from disorder, that reverses entropy, requires the expenditure of energy.

Now let's return to a chemical reaction. Many reactions require that two molecules come together for a reaction to take place. The majority of chemical reactions are performed in solution, meaning that the reactive molecules are dissolved in a liquid. These reactive molecules are swimming in a sea of inert (unreactive) solvent molecules. Somehow, they must find each other.

Two Boston Red Sox fans get tickets to a Yankees game in New York City. Unfortunately, their seats are nowhere near each other. They decide to meet up at Gate 4, the main front entrance to the stadium when the game ends. After the final out, each friend makes his way toward the exit. They are two Boston Red Sox jerseys among a sea of blue Yankees jerseys, bobbing and weaving through the crowds, pushed in this and that direction. After a great deal of effort fighting through the sea of people, the two friends bump into each other, seemingly accidentally, at the gate, pleased to have found each other after such madness in the crowd.

For many chemical reactions, a similar process has to occur. Two molecules have to wander among a sea of solvent molecules, randomly bumping into solvent molecule after solvent molecule, moving in one direction and then bumping into another direction. But no reaction can take place until these two molecules get close enough so that they can interact. Instead of being highly disordered, with the molecules far removed and moving about in random directions, the two molecules must be very ordered—they have to be right next to each other! For a reaction to proceed, the system needs to lose entropy and gain order. To lose that entropy and to order the system by bringing the reactant molecules near each other, energy must be provided.

The last consideration is that often it is not enough to simply bring the reactant molecules near each other. Usually, the molecules must be properly oriented to each other, bringing the reactive components together in the right way. Think of hammering a nail into a piece of wood. It is not sufficient to just hit the nail with the hammer. Rather, the point of the nail needs to be positioned at the face of the wood. The head of the nail needs to face out, and the hammer has to be brought down onto the nail head. How many times have we hit the nail with the hammer at a poor angle and bent the nail, or even worse, smashed our thumb? This proper orientation of the reactant molecules is again an ordering of the system, a further loss of entropy, with an energy requirement.

The upshot is that virtually every chemical step has an energy cost to overcome. This is called the *activation barrier*, the energy that must be supplied for the reaction to go. We can picture this energy barrier with the plot shown in Figure 3.5. The *x*-axis is the reaction coordinate. Think of this as representing the changes in the positions of the atoms in the molecules as the reaction takes place, analogous to the "straightened" route of the *Tour de France* stage shown in Figure 3.2. Moving from left to right along the *x*-axis means moving from reactant to product for the chemical step. We plot the energy change at each point along the reaction path on the *y*-axis. The resulting plot looks similar to a stage profile of a cycling race. The reaction begins with molecules

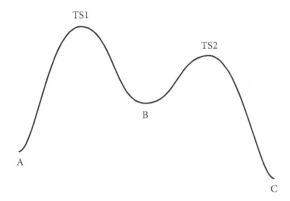

**Figure 3.5.** General reaction coordinate diagram.

at **A**. Energy needs to be supplied to climb the hill **TS1**, after which energy is released as the molecules get to **C**. Then a second barrier, a second hill must be climbed (**TS2**), and then it's all downhill, with energy being released until the molecules get to product **C**.

The minimum energy that must be supplied is the *activation energy* for that step. The point on the reaction coordinate that corresponds to the energy barrier is called the *transition state*. A single transition state is associated with every chemical step. In other words, a chemical step is the conversion of reactant to product passing through one and only one transition state. The reaction shown in Figure 3.5 takes reactants **A** through one chemical step to molecules **B** and then a second step produces the product **C**. We call molecules made along a multistep reaction *intermediates*, such as **B** in Figure 3.5.

You may have noticed that the activation energy is the minimum energy needed to get the reaction to proceed. It is certainly possible to supply even more energy and get over the barrier, just as one might decide to jump up right at the point of crossing the top of a mountain to get to the other side. The path from reactant to product through the transition state is called the *minimum energy path*, and other higher-energy paths can exist. But no path will take the molecule from reactant to product through a given transition state with a lower energy barrier.

Chemical steps can result in energy being produced or consumed. In both cases, an activation barrier must be crossed, meaning that energy must be supplied. These two cases are graphically represented in Figure 3.6. In the left plot, as the reactants **M1** change to products, the energy first increases until it reaches the transition state **TS(1)**. The energy gain is $E_a(1)$. After the transition state, the reaction proceeds to product **M2**. We use the notation $\Delta H$ to

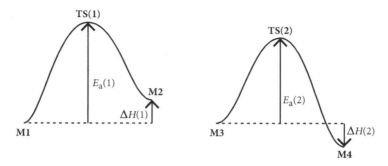

**Figure 3.6.** Reaction coordinate diagram for endothermic (left) and exothermic (right) reactions.

designate the change in enthalpy, where enthalpy ($H$) is a very close relative of energy. For the left reaction, the change in energy is positive, meaning that a net gain in energy took place (called an *endothermic reaction*). For the right plot, the reaction takes **M3** through **TS(2)** to the product **M4**. The activation energy is $E_a(2)$ for passing through **TS(2)**, but in this case the overall change in enthalpy $\Delta H(2)$ is negative, or a loss of enthalpy. This is called an *exothermic reaction*, where energy is released.

In general, exothermic reactions are preferable. Loss of energy is inherent in nature. A rock at the top of a hill will eventually make its way to the bottom because gravity pulls mass toward the center of the earth. On its own, a rock at the bottom of a hill will not move to the top. Rather, some external energy must be supplied to lift the rock up.

Of critical concern in developing a reaction mechanism are these energy changes. What is the size of the activation barrier? Is the reaction exothermic (releasing energy) or endothermic (consuming energy)? Is the reaction gaining or losing entropy? We will often seek answers to these energy questions when figuring out a reaction mechanism.

# 4

# Potential Energy Surface Features

In this chapter, I present a more detailed discussion of the important locations or points on a reaction coordinate diagram. In order to precisely define the nature of an intermediate or transition state, we would need to avail ourselves of calculus. Here, however, I will avoid the use of mathematics, instead invoking analogies to real-world experience to understand these concepts. Nonetheless, if you do find this chapter hard-going, it can be skipped without diminishing your ability to make it through the remainder of the book.

The reaction coordinate diagrams, such as those shown in Chapter 3, present just a slice through what is more fully considered as the *potential energy surface* (PES). We first need to talk about internal coordinates of a molecule. In our everyday lives, we are familiar with three dimensions or three coordinates. Simply put, we can move front or back, left or right, or up and down, and that's our three-dimensional space. With the widespread use of GPS, we might consider our position in terms of latitude, longitude, and altitude—again, three dimensions.

Now consider the locations of two people, say Alice and Bob. We can specify both Alice's and Bob's latitude, longitude, and altitude. However, all that might interest Alice is the distance to Bob, a single value. So, from just Alice's perspective, a single coordinate suffices to describe her relation to Bob. Obviously, the same argument holds for Bob.

Let's add a third person, Charlie—see Figure 4.1—standing on top of a hill. From Alice's perspective, there is the distance to Bob ($r_1$), and now there is also a distance to Charlie ($r_2$). But we need something to measure the distance between Bob and Charlie. The simple answer is the distance $r_3$. However, an equally valid measure is the angle α that lies between the lines connecting Alice and Bob and Alice and Charlie. Knowing the two distances $r_1$ and $r_2$ and the angle α and dredging up those spider-web-encrusted trigonometry lessons from high school will determine that distance between Bob and Charlie. The key here is that with three people, or three objects, or three atoms, we need three internal coordinates to describe their relative positions.

Adding in more people complicates matters further. With four people (or four atoms), as in Figure 4.2, we will require six coordinates, that is,

*Thinking Like a Physical Organic Chemist.* Steven M. Bachrach, Oxford University Press. © Oxford University Press 2023. DOI: 10.1093/oso/9780197640371.003.0004

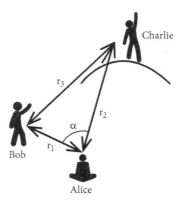

**Figure 4.1.** Three-person coordinates.

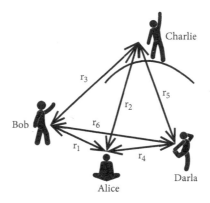

**Figure 4.2.** Four-person coordinates.

six measures to define the relationship of them. A system with five objects requires nine coordinates. This requirement can be generalized: the number of coordinates (or dimensions) is $3N$-6, where $N$ is the number of objects (or atoms).

Here is the simple way to think of that formula. Every atom, every object, can be defined by its $x$-coordinate, $y$-coordinate, and $z$-coordinate in our normal everyday three-dimensional space. That's the $3N$ part of the formula. Now think of a box, focusing on its corners (see Figure 4.3a). Moving the box to the left or right does not change the distances between the corners. Similarly, moving the box up or down or back and forth also doesn't change the box. So that's three coordinates that don't matter in defining the internal coordinates of the box. In other words, where the box is positioned

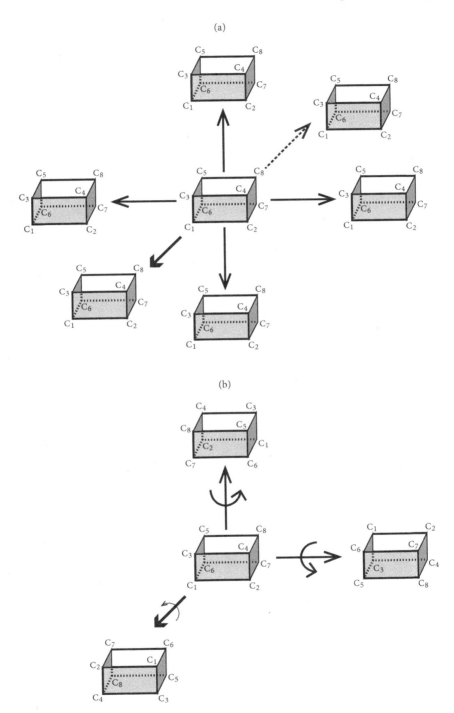

**Figure 4.3.** (a) translational coordinates and (b) rotational coordinates.

in our real three-dimensional space does not change the box. Similarly, you can rotate the box around the $x$-axis, $y$-axis, or $z$-axis (Figure 4.3b), and again the box itself doesn't change; the distances between the corners are unaffected. That removes another three coordinates, leaving $3N$-6 coordinates necessary to define the internal structure of the box, or of any object—including a molecule.

The potential energy surface displays the energy of a molecule with respect to each one of its coordinates. The Morse Potential shown in Figure 3.4 is the PES for a molecule containing two atoms, with the $x$-axis representing the single coordinate: the interatomic distance. (A careful reader might ask how a two-atom molecule has any coordinate, since applying the formula gets us $3*2 - 6 = 0$. The formula is modified to $3N$-5 for linear systems.) For larger molecules, this PES becomes difficult to draw as the dimensionality gets very large with the increasing size of the molecule. For a three-atom molecule, such as water, $H_2O$, we will need to plot the energy against three coordinates (the distance from oxygen to the first hydrogen, the distance from oxygen to the second hydrogen, and the angle between the two O-H bonds), resulting in a four-dimensional plot—which is difficult to present on just a two-dimensional sheet of paper! This is why we so often use the reaction coordinate diagram, which reduces this high-dimension plot to just two dimensions: the energy along the $y$-axis and the reaction coordinate (the path that takes us from reactant to product as discussed in the previous chapter) on the $x$-axis.

(a)

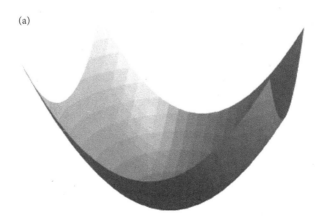

**Figure 4.4.** Model potential energy surfaces.

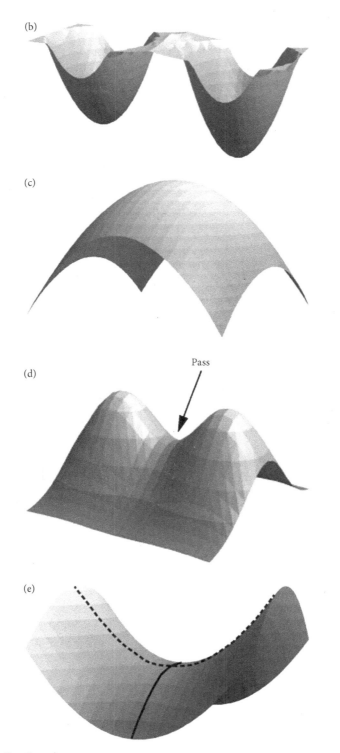

(b)

(c)

Pass

(d)

(e)

**Figure 4.4.** Continued.

Let's now consider a few examples of features on PESs. For these examples, let's consider a PES for a molecule that has two independent coordinates, and on the third axis we'll consider the energy associated of that molecule at every value of the pair of those two coordinates. Our first case is shown in Figure 4.4a. It looks like a crater, and let's consider what the world looks like in a crater. A ball placed anywhere on the surface of the crater will start to roll downward and will eventually settle to rest right at the bottom of the crater. Why? Gravity pulls object toward the center of the earth because that's where the energy is lower. The bottom of the crater is the closest approach to the center of earth: it's the lowest energy position for the ball. Moving the ball a small amount in any direction—left or right, front or back, takes the ball into a higher position, having more energy. Let go and the ball will roll right back. The bottom is the energy minimum and is a stable point on the surface. Any movement from that position leads to a higher energy.

This bottom of the crater is where stable molecules exist. The arrangement of the atoms is such that the molecule sits in a bottom of a $3N$-6 well. Any small change in the position of any atom in the molecule will lead to a higher-energy arrangement.

Notice that I was being very careful in saying "any *small* change in the position of any atom." Consider the PES in Figure 4.4b. The surface has two wells. If you sit at the bottom of the left well, any small displacement left or right or front or back leads to an increase in energy. However, if you climb up out of the well moving to the right, you can then fall into the second well, which is actually lower in energy. Once you are at the bottom of this right well, again any small movement leads to an increase in energy. We use the term *local energy minimum* to refer to the point at the bottom of a well on a potential energy surface. Calculus provides formulas for describing such points. What about finding the lowest energy point over the entire PES, the *global energy minimum*? For example, suppose we are in the left well; can we know that there is an even lower well to our right? The answer is that calculus provides no such way of knowing the location of the global energy minimum. To identify the global minimum, we would have to explore the *entire energy surface*, keeping track of the energy at the bottom of each well, every local energy minimum, and then select the lowest one of this collection. Molecules that are reactants, intermediates, or products can be characterized as local energy minima.

Let's next examine Figure 4.4c, which is the opposite of a well. Here, we see a hilltop, where any movement will lead us to tumble downhill. This is a point of instability, as any change in position will lead to a lower-energy structure. Hilltops tend to be avoided, and they have no importance in chemistry.

How can hilltops be avoided? Unless one is a mountain climber or one of those crazy cyclists in the Tour de France, usually we will be interested in getting to the other side of the mountain and would rather not have to expend the energy to climb all the way to the top. A great solution is a tunnel, and we will actually discuss that option in a later chapter. In most cases, what we want is to find a pass between mountains, that low point that separates two peaks. Figure 4.4d shows two hills with a pass between them. The early pioneers leading a wagon train over the Rockies instinctively knew to cross at the pass, and not by a route that went higher up the hills.

Let's take a closer look at the region around the pass, as in Figure 4.4e. This looks like a mash-up of a well and a hill. The dashed line defines a well, with the lowest point at the pass, whereas the solid line defines a hill, with the highest point at the pass. In one direction, given by the dashed line, we sit at the bottom of a well, while in the perpendicular direction, the solid line, we sit at the top of a hill. To travel from one side to the other, we climb up the solid line until we hit the pass, the intersection of the dashed and solid lines.

The pass defines the minimum energy needed to cross over to the other side. One can certainly take another path, say, swinging to the right of the pass and climbing higher up the right peak before descending on the other side. This alternative does take you over to the other side, but you have to expend more energy going this route.

The solid path is the most direct, least-energy path that takes one up and over the barrier. It is the reaction path from in front of the range to the other side, and the pass is that top of the barrier. This solid path looks just like the plots shown in Figure 3.6! Connecting back to chemistry, we see that the reaction path starts with reactants, goes up in energy through the transition state, and then down in energy to product. The *transition state* is where the molecule sits in the bottom of the well for all of the coordinates except for one coordinate—and that's the reaction coordinate. Along that coordinate, the transition state sits at the top of the hill. Figure 3.6 and any typical reaction coordinate diagram only portrays the coordinate for the hill all of the other coordinates, which look like wells, are omitted from the diagram.

To summarize, a chemical reaction starts with a reactant molecule at its local energy minima. As the distances and angles (i.e., the internal coordinates) between the atoms start to change, the energy rises until the transition state is reached, the point where the molecule sits in a well for all coordinates but one; in that one coordinate, the molecule is at the top of a hill. Further progress along the reaction coordinate leads to lowering the energy until the product is reached, which is again a local energy minimum.

# 5

# Nucleophilic Substitution Reactions.
# I. Basics and S$_N$1

The first few weeks of a typical sophomore organic chemistry course covers definitions, chemical bonding, and naming conventions. It's all pretty straightforward stuff, often lulling students into an unproductive sense of complacency. What comes next is often a very rude awakening for many students as the first organic reaction involves two competing reaction mechanisms. For most students, understanding this reaction requires a new way of thinking.

I vividly remember this period of my own sophomore year at university. I could not make heads or tails of these mechanisms. There didn't seem to be any logic to it, and I simply did not want to go through organic chemistry relying on sheer memorization. I was beginning to question whether I should remain a chemistry major. If I couldn't manage my second year of chemistry class, what was going to happen in subsequent classes? It did not bode well, I thought.

I remember I came home for a weekend. One night I was up around midnight studying organic chemistry, and my father walked into my room on one of his late-night wanderings. Now my father was a professor of chemistry and his area was organic chemistry. He had taught this course at least 20 times. He had certainly seen countless students struggle with this reaction.

My father sat with me, and together we talked about the reaction and why the reagents were coming together to react. He was able to provide a logical structure that I could incorporate into my own thinking. I vividly remember feeling like a light bulb had suddenly switched on in my brain at that moment. I felt like this all made sense, that chemicals were behaving according to sensible rules, and that these rules would allow me to predict what would happen in related situations. I felt that I had a structural framework on which to build out a full understanding of organic reactions. I certainly also felt relieved and became convinced that this semester would be OK. And in fact, it was; that does not mean that the rest of organic chemistry was a breeze! I still worked very hard, but I never again felt adrift.

*Thinking Like a Physical Organic Chemist*. Steven M. Bachrach, Oxford University Press. © Oxford University Press 2023.
DOI: 10.1093/oso/9780197640371.003.0005

So what is this first organic reaction that caused me so much angst and that plagues so many students? A simple example of this reaction category is the first reaction I presented in Chapter 2:

$$\underset{\underset{H}{|}}{\overset{\overset{H}{|}}{H-C-Cl}} + H-Br \longrightarrow \underset{\underset{H}{|}}{\overset{\overset{H}{|}}{H-C-Br}} + H-Cl$$

On the face of it, this reaction appears so easy—we've simply swapped the chlorine and bromine atoms. In fact, that's where the name comes from: these reactions are called *substitution reactions*.

Let's go a little further in specifying its name. Remember that every atom wants to achieve the same number of valence electrons as the closest noble gas element, those elements in the rightmost column on the periodic table. Chlorine and bromine are in the column adjacent to the noble gases, and this group is called the *halogens*. They both would like to pick up one electron, and if they do that, they will carry a charge of –1. In HBr, the bromine can take an electron from hydrogen, and that would leave hydrogen with no electrons, a situation that is actually quite common. Bromine with a negative charge ($Br^-$), called bromide, would be attracted to anything with a positive charge. Recall that the positive charge within an atom is concentrated at the nucleus, so $Br^-$ would be attracted to a nucleus. Turning to the Greek language, we can describe $Br^-$ as being a *nucleophile*, a lover of nuclei. All negatively charged atoms or molecules are nucleophiles, and even some neutral species will behave as nucleophiles. Thus, our prototype example above can be called a *nucleophilic substitution reaction*.

To start our discussion of nucleophilic substitution reactions, let's write up a generic formulation of the reaction:

$$\underset{\underset{R3}{/}}{\overset{\overset{R1}{\backslash}}{R2-C-LG}} + Nuc^- \longrightarrow \underset{\underset{R3}{/}}{\overset{\overset{R1}{\backslash}}{R2-C-Nuc}} + LG^-$$

The reactant $Nuc^-$ is the nucleophile, and it will replace the leaving group, appropriately labeled $LG^-$. The leaving group can be a single atom, like the chloride in the example above, or a collection of atoms that will leave together as a group. In addition to the leaving group, carbon will have three other groups bonded to it, which I have labeled R1, R2, and R3. These can be composed of a single atom or groups of atoms. These three groups can be identical, like the three hydrogen atoms in the first example in this chapter, or they can be different. Importantly, the three R groups remain unchanged during the course of the reaction: they start out connected to carbon and remain connected to carbon, and no changes occur to them.

We can now examine our generic nucleophilic substitution reaction for the bond changes that are taking place. The C–LG bond will be broken. The C–Nuc bond will be made. That's it: one bond broken and one bond made. What could be simpler?

Let's imagine all the possible itineraries, if you will, for these two bond changes. There are three cases.

| | |
|---|---|
| Case 1 | Step 1: Make the C–Nuc bond. |
| | Step 2: Break the C–LG bond. |
| Case 2 | Step 1: Break the C–LG bond. |
| | Step 2: Make the C–Nuc bond. |
| Case 3 | Step 1: Accomplish both changes in the same step. In other words as the C–LG bond is breaking, start making the C–Nuc bond. |

For case 1, the reaction mechanism would be as follows:

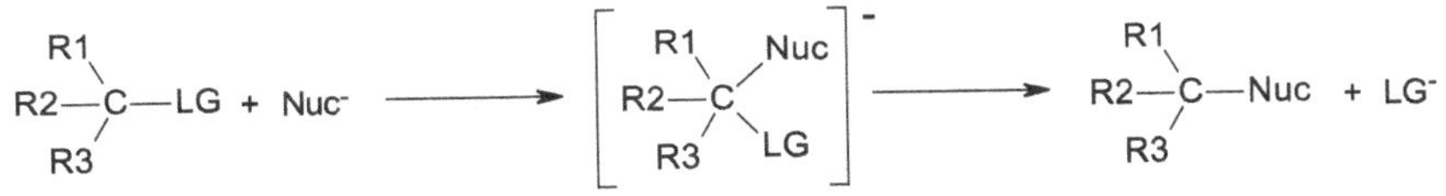

Let's take a close look at the intermediate on this proposed path. The central carbon atom makes five bonds, one to each of the original three R groups, the fourth to the leaving group that still remains, and the fifth bond is the new one to the nucleophile. That makes 10 electrons around carbon, and that's pretty much the cardinal sin for an organic chemistry student to make in class. Every carbon atom is desperate for a filled octet, and though we occasionally will observe a carbon with six or seven electrons, going beyond the octet does not happen. This mechanism requires making an intermediate that is so incredibly unstable that it has not been observed. That means we can completely discount Case 1 as a possible reaction mechanism to explain nucleophilic substitution reactions.

Well, that was certainly easy and encouraging! However, the other two mechanisms cannot be so readily treated. Let's walk through what each reaction pathway looks like. For Case 2, the mechanism is

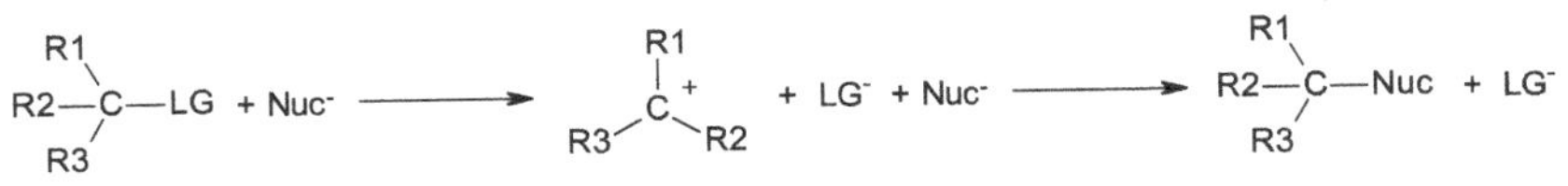

In the first step, the one indicated by the left arrow, the C–LG bond is cleaved, creating the free leaving group LG⁻ so that what is left behind must have a positive charge. The positive charge resides on the central carbon atom, and that makes sense given that this carbon atom does not have a full octet. It has only six electrons. The two electrons that had constituted the C–LG bond took off with the leaving group. We call a carbon atom with a positive charge a *carbocation*. We might expect carbocations to be fairly reactive species, due to carbon's unfilled octet and its positive charge. In fact, carbocations are reactive species, but they do exist and they have been observed as stable, though short-lived, species. It is important to note that the nucleophile is not involved at all in this first step. It is present, but it's a bystander in this step. The second step brings the carbocation and the nucleophile together to make the new C–Nuc bond. The resulting product again has the carbon with a complete octet.

In total, both of these steps appear to be chemically reasonable. It's likely that the first step is more difficult and has a larger activation barrier than the second step since the C–LG bond must be fully cleaved. The second step appears to have a relatively small barrier as two oppositely charged species combine, producing an electrostatically favorable outcome.

How about Case 3, where the bond breaking and bond making take place in a single step? This mechanism can be depicted as

$$
\begin{array}{c}
\text{R1} \\
\text{R2}\text{—}\overset{\displaystyle |}{\underset{\displaystyle /}{\text{C}}}\text{—LG} + \text{Nuc}^-
\end{array}
\longrightarrow
\left[\begin{array}{c}
\text{R1} \\
\text{Nuc} \text{---} \overset{\displaystyle |}{\underset{\displaystyle /\backslash}{\text{C}}} \text{----LG} \\
\text{R2}\quad\text{R3}
\end{array}\right]^{\ddagger -}
\longrightarrow
\begin{array}{c}
\text{R1} \\
\text{R2}\text{—}\overset{\displaystyle |}{\underset{\displaystyle /}{\text{C}}}\text{—Nuc} + \text{LG}^-
\end{array}
$$

That piece in the middle, contained in the square brackets, needs some explanation. The double dagger at the top right corner indicates that this species is a *transition state* and <u>not</u> an intermediate. The dashed line indicates a partial bond, so this depiction indicates a partly formed C–Nuc bond and a partly broken C–LG bond. The entire reaction takes place in a single step, with the top of the energy barrier, the transition state, having these two bonds only partly formed. Since the bonds are partly formed, the octet rule is not broken; one can think of this as, say, half a C–Nuc bond and half a C–LG bond, together accounting for two electrons across these two partial bonds. The extent to which the C–LG has broken and the C–Nuc bond has formed is open to debate and specific circumstance. At this point, just don't get stuck thinking that it has to be half and half.

OK, we have two seemingly viable reaction mechanisms: Case 2, which has a two-step process, and Case 3, which has a single step. How do we decide which one, if either, actually takes place? Here's where experiment is critical.

We need some data that can help discriminate between the two mechanisms. The first thing we will look it is *kinetics*, the rate of a reaction.

Kinetics is the study of rates, determining how fast something occurs. This can be the rate of travel of a car, in miles per hour, or how fast a heart beats, in beats per minute, or how fast the electricity oscillates in household current, in cycles per second. As for chemists, we are interested in how fast a reactant disappears or how quickly a product is made. So, we might measure the amount of a reactant we place in a flask and then the amount that remains in 30-second intervals. The ratio of how much reactant is lost per time period is the rate of the reaction. Similarly, we could monitor how much product is present at various time steps—zero at the start and increasing with time—and get the rate of the reaction.

An example of what experimental data looks like for a kinetics experiment is given in Figure 5.1. The *x*-axis is time, and on the *y*-axis we've plotted how much reactant (circles) and product (squares) are present. At the start, only reactant is present, but over time the amount of reactant diminishes and the amount of product grows. You may notice that the reaction does not head to a situation where all the reactant is gone and only product remains.

Most of the time we will be interested in the rate of reaction right at the start. Looking at the product curve, the squares, we note that the product grows in rapidly at the beginning (a sharp rise in how much product is present). The reaction slows down as time progresses, reaching essentially zero change at large times (note that the amount of product stops increasing at the far right of the plot). If you recall some calculus, computing the slope of the curve at the start will provide the initial rate.

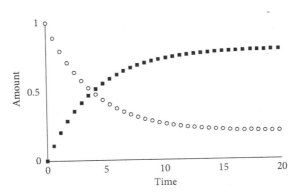

**Figure 5.1.** Kinetics experiment results. Circles are concentration of reactant diminishing with time, and squares are concentration of product increasing with time.

Suppose that we are studying a reaction involving two different reagents, say A and B, and that these combine to create product C, A + B → C. What happens to the initial rate if we double the amount of A? Or double the amount of B? Or double both reagents? I think our intuition would tell us that doubling A would speed up the reaction, that having more of either reactant means it's more likely for A and B to find each other and react. Perhaps doubling the amount of A might double the rate, as would doubling the amount of B, and doubling both at the same time might increase the rate by a factor of four.

(As an aside, we most often conduct these types of rate experiments in solution. We would then measure the amount of a reagent as a concentration. We denote a concentration of a reagent as the name of the reagent within square brackets, so the concentration of A is written as [A].)

Let's perform this set of experiments on the following reaction:

$$
\underset{\substack{\displaystyle | \\ CH_3}}{\overset{\substack{CH_3 \\ \displaystyle |}}{H_3C-\overset{|}{C}-Cl}} + \underset{\substack{\displaystyle | \\ H}}{\overset{\substack{H \\ \displaystyle |}}{H_3C-\overset{|}{C}-O-H}} \longrightarrow \underset{\substack{\displaystyle | \\ CH_3}}{\overset{\substack{CH_3 \\ \displaystyle |}}{H_3C-\overset{|}{C}-O}}\underset{\substack{\displaystyle | \\ H}}{\overset{\substack{H \\ \displaystyle |}}{\overset{|}{C}-CH_3}} + \quad H-Cl
$$

(To simplify the notation, the hydrogens bonded to a carbon are simply noted next to C, such as $CH_3$.) We will measure the initial rate of this reaction and then repeat the reaction with twice the concentration of the first reagent, called *tert*-butylchloride. Our third experiment doubles the concentration of the second reagent, called ethanol, and the last experiment doubles the concentration of both reagents. Our finding is perhaps not what was expected. Doubling the concentration of *tert*-butylchloride does double the rate of the reaction. However, doubling the concentration of ethanol does not affect the rate at all. The last experiment, with twice the concentration of both reagents, has a rate that is only twice as fast as the first experiment. Even though ethanol is a reagent—it *does disappear during the course of the reaction*—it has no effect on the rate!

How is this possible? How can the amount of a reagent present in the flask *not affect the rate of the reaction*? To understand this problem, we need to once again consider potential energy surfaces. This time we imagine a roller coaster, one that might look like the track shown in Figure 5.2.

The ride starts at the left-hand side, with a cable that pulls the car up that first hill. In pulling the car and the people inside it, the cable has a lot of hard work to do, fighting the force of gravity the entire way up. Once the car hits the top of this first hill, it disengages from the cable and speeds along downhill without any further assistance. The car passes through the loop and then climbs back uphill once more, but never again does it reach a height as tall as

**Figure 5.2.** Diagram of a roller coaster ride starting on the left.

that initial hill. The potential energy available at the top of the first hill, provided by the cable pulling the car up, is converted to kinetic energy as the car zips down the track.

Now think about the time and energy needed to do the climbing. Three hills must be crossed, creating three expenditures of energy that must be paid. Clearly, it's that first hill that is the toughest—in every regard. The cable has to drag everyone up that tall hill. It takes quite some time inching its way upwards. But the rest of the ride seems to be completed in the blink of an eye. Although the car does slow down somewhat when it climbs through the loop and over the last hill, the slowest part certainly occurred when it climbed the first hill.

Now consider the reaction potential energy surface shown in Figure 5.3. This is a two-step reaction, with the activation barrier much higher for the first step than for the second. As was the case for the roller coaster, it will take more energy and more time to cross that first barrier. In fact, once that first barrier is crossed, the reaction will usually have enough energy to fall downhill and shoot across the second barrier. That first, higher barrier dictates how quickly a reaction will take place. We call that step the *rate-determining step*, the step that involves crossing the highest energy barrier. It doesn't have to be the first step; if we have a multistep reaction, whichever barrier is the highest will be the rate-determining step. The overall reaction can go no faster than the rate to cross the highest barrier, the rate-determining step. In our roller coaster example, crossing over that first hill is its rate-determining step. (To be fair, at most amusement parks, the actual rate-determining step will be the time spent standing in the queue waiting to get on the ride.)

The kinetics of a reaction that has a surface like that shown in Figure 5.3 will be dictated by the rate-determining step for short reaction times. As the reaction goes on, the kinetics becomes more complex. If the energy difference

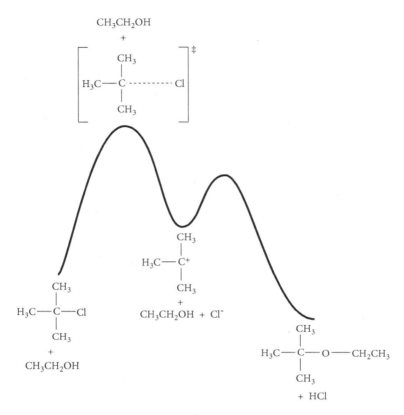

**Figure 5.3.** Reaction coordinate diagram for an $S_N1$ reaction.

between the activation barrier of the rate-determining step and any other step is fairly small, this simple model will also break down. For our purposes here, our simplified model works well.

Returning now to our nucleophilic reaction involving *tert*-butylchloride and ethanol, we can use the sketch in Figure 5.3 to model its potential energy surface. We have already argued that the first step, breaking the C–Cl bond, will have a larger barrier than that in the second step, making the C–O bond. I have added some drawings to Figure 5.3 to label some important points. Most notable is the transition state of the first step. This first step is the rate-determining step; it has the higher activation barrier. Now what is happening in this rate-determining step? Simply the breaking of the C–Cl bond. The nucleophile, ethanol, the other reagent, is not involved in this bond breaking, so it is not surprising that the amount of ethanol present has no effect on the rate that the C–Cl bond breaks. Ethanol is a bystander, watching that bond break, and only then can it get in on the action and attack the carbocation.

That attack, however, will be much faster than the bond breaking of the first step, so it does not affect the overall reaction rate.

We can write a rate expression for this reaction as $rate = k[tert\text{-}butylchloride]$, where $k$ is the rate constant, determined by experimental measurement. The important part of this expression is that only the concentration of *tert*-butylchloride appears in the expression and no other reagent. We term this a first-order reaction, since the concentration of only one substance appears in the rate expression and that concentration is raised to the first power. We can also say that the rate-determining step is *unimolecular*, involving just a single molecule.

What else might we test about this mechanism? The difficult step to take is that first one, forming a carbocation. Would making some small changes that might stabilize or destabilize the carbocation affect the rate of the reaction? Whatever changes we might make to the compounds would need to be relatively small so as to not change the reaction mechanism.

An organic chemist's first thought with regard to making small changes would be to alter the degree of substitution at the carbon that carries the leaving group. In Figure 5.4, I show four examples of differing substitution. Look closely at the carbon that is bonded to chlorine. In the first molecule, chloromethane, beside the C–Cl bond, the three other bonds are to different hydrogen atoms. In the second molecule, one of the C–H bonds is replaced with a C–C bond. Moving further right, we continue that change, replacing a C–H bond with a C–C bond. We think of the C–C bond as being the baseline, the simplest component, of an organic molecule. Moving left to right means adding more C–C bonds, but these differences should be minor perturbations, not a big change such as adding another element or double bonds. Organic chemists name this substitution pattern primary (1°), secondary (2°), and tertiary (3°), indicating the number of C–C bonds at the carbon with (in this case) the leaving group.

Why might it be useful to examine this series of compounds? Well, the stability of carbocations increases with substitution. In other words, as shown in

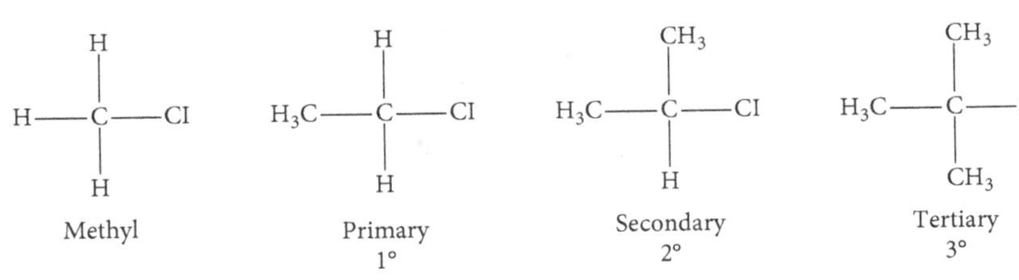

Figure 5.4. Definition of substitution degree.

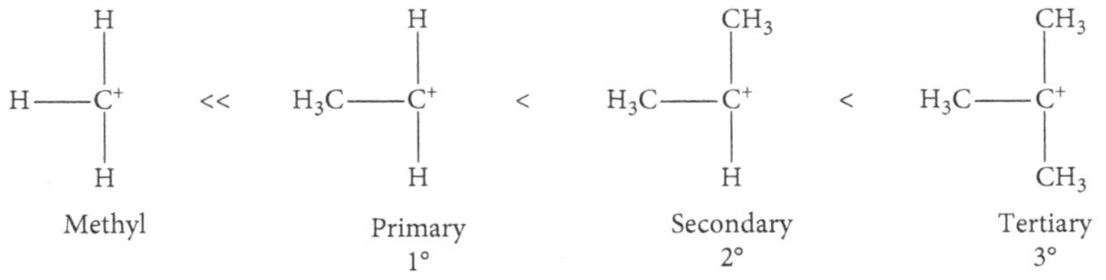

**Figure 5.5.** Relative stability of carbocations.

Figure 5.5, the most stable carbocations are the tertiary ones. The secondary carbocations are somewhat less stable than the tertiary carbocations. Primary carbocations are significantly less stable than the secondary examples. And the methyl cation is extremely unstable, being extraordinarily difficult to identify at all.

That might imply that cleaving the C–LG bond will be easier in a tertiary system than in a secondary system. That should lead to nucleophilic substitution reactions where tertiary systems react faster than with secondary systems, which should react much faster than primary systems. This trend has been observed many times, with different leaving groups, nucleophiles, and organic substrates.

The British organic chemist Sir Christopher Ingold coined the name $S_N1$ for this reaction mechanism. It stands for substitution (S), nucleophilic (N), and first order (1). Reactions that follow the $S_N1$ mechanism express first-order kinetics, and tertiary systems react faster than secondary systems. Primary systems do not readily react at all.

It's not the only way that a nucleophile can replace a leaving group, and that's Case 3, the subject of the next chapter.

# 6

# Nucleophilic Substitution Reactions. II. S$_N$2

The reaction of iodomethane with cyanide proceeds rapidly, with the substitution of cyanide for iodine:

$$H_3C{-}I\ +\ ^-C{\equiv}N\ \longrightarrow\ H_3C{-}C{\equiv}N\ +\ I^-$$

This statement seems to violate what we discussed in Chapter 5, the idea that a reaction that would invoke a methyl cation could take place. The fact that this reaction does occur and that innumerable nucleophilic substitution reactions of methyl and primary substrates are known, suggests that a second, alternative, reaction mechanism, fundamentally different from the S$_N$1 reaction, must occur. We leave it to a later chapter to discuss what reaction conditions are needed to experimentally observe either type of reaction mechanism. For now, we'll gloss over this aspect but delve into some of the experimental observations of this second type of nucleophilic substitution reaction.

Let's begin with kinetics. Using our reaction above, we will again run a set of four experiments measuring the initial rate of reaction. The first experiment has an equal amount of each reagent, iodomethane and cyanide. In the second experiment, we'll double the concentration of iodomethane, while keeping the original concentration of cyanide. The third experiment swaps this change, doubling the concentration of cyanide and using the same concentration of iodomethane as in the first experiment. In the fourth experiment, we'll double the concentration of both reagents.

This time, the results seem more intuitive. Doubling the concentration of either reagent (experiment 2 or 3) results in a reaction rate that is twice as fast. Doubling the concentration of both reagents (experiment 4) results in a reaction rate that is four times as fast. We can write a rate expression for this situation as $rate = k[\text{iodomethane}][\text{cyanide}]$. Here, the overall rate is second-order, first-order iniodomethane and first-order in cyanide. This result indicates that the reaction is *bimolecular* in the rate-determining step, involving one molecule of each reagent.

*Thinking Like a Physical Organic Chemist.* Steven M. Bachrach, Oxford University Press. © Oxford University Press 2023. DOI: 10.1093/oso/9780197640371.003.0006

A bimolecular reaction implicates the formation of the C-nucleophile bond in that slow step. But that can't be the only bond change that occurs in this step; if it were the only change, then we would be operating under Case 1. Recall that we removed Case 1 from consideration: forming the C-nucleophile bond first would mean having a carbon with five bonds and ten electrons, and we don't ever have that!

How can we avoid creating a carbon with five bonds? Well, Case 3 shows us a way. As the new C–Nuc bond is being formed, the C–LG breaks. At no specific moment would the total bonding exceed eight electrons. This is a *concerted* process where, within a single step, multiple bonding changes take place.

The reaction coordinate diagram for this reaction is shown in Figure 6.1. The nucleophile (cyanide in this case) approaches the carbon atom and starts to form the bond. At the same time, the leaving group (iodide in this case) begins to exit. The energy increases until the transition state is reached, with a partially made C–CN bond and a partially broken C–I bond. After the transition state, the C–CN bond continues to form, while the C–I continues to break, with a smooth decrease in energy, until the final products are formed. The reaction has a single step, with a single activation barrier. Analogous to the name of the other reaction mechanism, Christopher Ingold named this $S_N2$ for nucleophilic substitution bimolecular.

It is important to recognize the distinctly different reaction coordinate diagrams for the $S_N1$ (Figure 5.3) and $S_N2$ (Figure 6.1) reactions. The $S_N1$ reaction is characterized by two distinct chemical steps (two transition states),

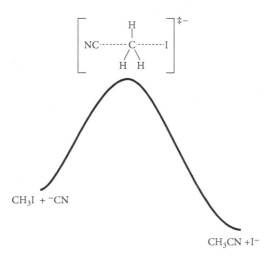

**Figure 6.1.** Potential energy surface for the $S_N2$ reaction of $CH_3I + CN^-$.

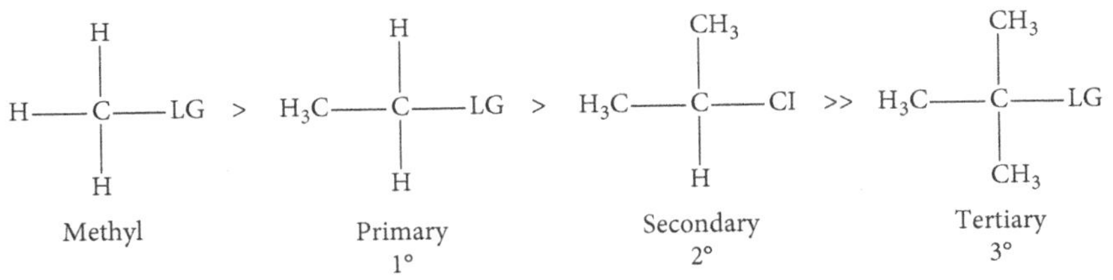

**Figure 6.2.** Relative reactivity of alkyl molecules in the $S_N2$ reaction.

a slow first step and a faster second step, while the $S_N2$ reaction is a single step (one transition state).

How does the degree of substitution at the reacting carbon atom affect the $S_N2$ reaction? Recall that for the $S_N1$ reaction, the more substituted, the faster the reaction; that is, 3° react much faster than 2° and 1° essentially don't react. Experiments reveal that the $S_N2$ reaction inverts this order (see Figure 6.2). For the $S_N2$ reaction, methyl systems react very fast, as do primary systems. Secondary systems are much slower, and tertiary systems essentially don't react.

To discuss why this trend is observed for the $S_N2$ reaction, we need to develop an understanding of the three-dimensional shape of molecules. I tackle that topic in the next chapter.

# 7

# Nucleophilic Substitution Reactions.
# III. Stereochemistry

What do molecules look like? When we talk about molecules coming together to react, it stands to reason that their shapes will matter. Atoms that are on separate molecules that will end up bonded to each other when the reaction is finished have to be able to find each other and get close. Imagine an atom buried inside a molecule; that atom is unlikely to break off in order to bond to an atom in another molecule. Conversely, an atom at the surface should be available to make a new bond. We might suspect that an atom forced into some unusual geometry might be susceptible to making a change, breaking and forming new bonds to get to a more usual arrangement.

The development of the understanding of an organic molecule's shape dates back to 1874 when the Dutch chemist Jacobus Henricus van't Hoff Jr., later to be the first recipient of the Nobel Prize in Chemistry, and the French chemist Joseph Achille Le Bel independently proposed the idea of a tetrahedral arrangement about a carbon atom. Their arguments were attempts to explain an important concept in organic chemistry, one that is essential to understanding nucleophilic substitution reactions, let alone life itself! I'll get to this specific concept in just a bit.

A tetrahedron is one of the Platonic solids, that is, three-dimensional objects whose faces are identical polygons. Probably the most familiar Platonic solid is the cube, constructed of six identical faces, each of which is a square. The tetrahedron is constructed of four faces, each of which is an equilateral triangle. The caltrop shown in Figure 7.1 has four sharp points pointing into the corners of a tetrahedron. No matter how it lands on the ground, one of these sharp points will be directed upward. (Police may scatter caltrops on a road to puncture the tires of a car driven by a suspected criminal.)

Here's an important geometry problem: how can you place four objects such that each is equally far apart from the others? The naïve first attempt might be a square (Figure 7.2a). The four sides are of equal distance, but the A–C and B–D distances are much longer than the sides. Perhaps a diamond might work better (Figure 7.2b); all of these sides are of equal length, but the A–D distance

*Thinking Like a Physical Organic Chemist.* Steven M. Bachrach, Oxford University Press. © Oxford University Press 2023.
DOI: 10.1093/oso/9780197640371.003.0007

**Figure 7.1.** Caltrop, with a tetrahedral shape, is used by police to puncture tires.

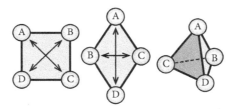

**Figure 7.2.** Placing four objects equidistant: (a) in a square, (b) in a diamond, (c) in a tetrahedron.

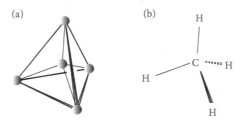

**Figure 7.3.** (a) Tetrahedral arrangement of four balls with a fifth ball in the middle and (b) the tetrahedral geometry of methane.

is longer. One has to bend one of these triangles of the diamond out of the plane to form the tetrahedron (Figure 7.2c), which solves our problem.

Imagine placing a ball at the center of the tetrahedron and a ball at each of the four corners, as shown in Figure 7.3a. The distances from the center ball to each corner ball are identical, indicated by the lines between the balls. Imagine if these lines represent the bonding pair of electrons for each of the four C–H bonds in methane, $CH_4$. Those pairs of electrons would be as far from each other as possible while still remaining near the central carbon atom. Since electrons repel each other, this tetrahedral arrangement of the bonding electron pairs would minimize their repulsion. This suggests that methane would have a tetrahedral structure (Figure 7.3b). The American physical chemist G. N. Lewis, the same person who developed the electron dot model, made this argument in proposing the rationale for the tetrahedral geometry about a carbon atom making four bonds to four different atoms.

I want to expressly point out the notation used in Figure 7.3b. The tetrahedral methane molecule is a three-dimensional object. A piece of paper or a computer screen is a two-dimensional object and is thus a limited medium for displaying three-dimensional objects. Chemists use the convention of a bold wedge to indicate a projection in front of the page or screen, and dashes or a dashed wedge to indicate a projection behind the page or screen.

Let's now return to the reaction of iodomethane with cyanide discussed in the previous chapter. The nucleophile, cyanide, has to make its way to the carbon atom to form the new C–C bond. We can draw iodomethane as seen in Figure 7.4, with the leaving group, iodine, pointed to the right and one hydrogen pointed upward and a little to the left. The tetrahedral geometry at carbon means that the other two hydrogens point down and to the left, with one coming in front of the page and one behind the page.

Iodine is much larger than any other atom here, and its size is represented by the circle centered on the atom. It is the leaving group, and it will exit to the right, as indicated by the curved arrow below it. If the cyanide were to attack from the right, the same side as the leaving group, it would have to somehow get around the big iodine to approach the carbon atom. Furthermore, both the incoming group, cyanide, and the exiting group, iodide, have some negative charge and will repel each other. This frontside pathway, nucleophile attack from the same side as the leaving group seems to present a difficult prospect.

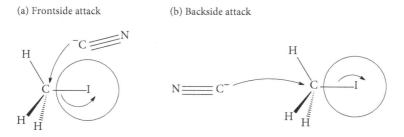

**Figure 7.4.** Mechanism for an $S_N2$ reaction via (a) frontside attack and (b) backside attack.

An alternative pathway might be to attack from the opposite side, or what we call a *backside attack*. The cyanide has to make its way between the very small hydrogen atoms. If it attacks the carbon along the line formed by the carbon and iodine, but on the backside, the path is unencumbered. As cyanide, the nucleophile, comes in from the left, the iodine can readily leave to the right, and neither the nucleophile nor leaving groups would get in each other's way. This backside attack is shown in Figure 7.4b, with the curved arrow from cyanide to carbon showing the new bond being formed, and the curved arrow on the right showing the breaking C–I bond. I'll have more to say about these curved arrows in a later chapter.

Let's see how a backside attack helps us understand the substitution trends in $S_N2$ reactions, shown in Figure 6.2. Attack at methyl and primary carbons from the backside seems quite reasonable. In the methyl system, as noted above, the back of the carbon has only small hydrogen atoms. In the primary system, there are two hydrogens, so even if the third group attached to the carbon is pretty big, the nucleophile can dip away from the large group toward the small hydrogen atoms and find its way to carbon. With a secondary system, there are only one small hydrogen atom and two large groups, and it's quite reasonable that the rate will be slower than for the primary systems as the pathway to the carbon is impeded. Lastly, a tertiary system has three large groups on the backside, and there's just no reasonable access between them to the carbon.

While this explanation seems quite rational, can we really nail down a backside attack with some strong evidence? The answer is definitely yes, but once again I need to make a bit of a tangential excursion here. I need to develop the concept of enantiomers, which is the notion that led van't Hoff and Le Bel to the tetrahedral carbon atom in the first place.

The simplest way to approach enantiomers is to look at your hands (Figure 7.5). Your left and your right hands are identical in most respects; each has a

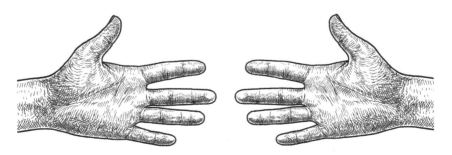

**Figure 7.5.** Mirror image of two hands.

palm and a backside, as well as five fingers from thumb, pointer finger, middle finger, fourth finger, and pinkie, in that order. But your two hands differ in one important respect. Position your left hand in front of you with the palm pointing away from you. Then place your right hand in front of the left such that the palm is pointing away from you as well. You will see that your palms and the back of your hands are in the same orientation (see Figure 7.6). Notice your fingers: each thumb lines up not with the other thumb but with the other pinkie! Your hands are not truly identical, but they are related.

Your hands are related by being mirror images of each other. Look at your right hand in the mirror, and what do you see? Your left hand, and vice versa. If we imagine a mirror coming out of the page between the two hands in Figure 7.5, then the hand on the left is reflected through that perpendicular mirror into the hand on the right. Your left and right hands are identical only as mirror images of each other.

An object that is not superimposable upon its mirror image, like each hand, is called *chiral*. Your left hand is chiral, as is your right hand. Your left foot and right foot are each chiral. You cannot superimpose your left foot onto your right foot.

Chirality has consequences that we are all familiar with, particularly when two chiral objects interact with each other. You can easily put a right-handed glove onto your right hand, but a left-handed glove will not fit on your right hand. The left shoe works just fine on your left foot, but good luck walking with the left shoe on your right foot!

*Isomers* are molecules that have something in common, like the same number of atoms of each element. *Stereoisomers* are molecules that are

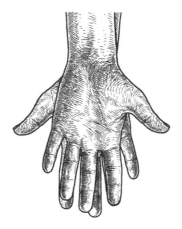

**Figure 7.6.** Attempting to match up two hands.

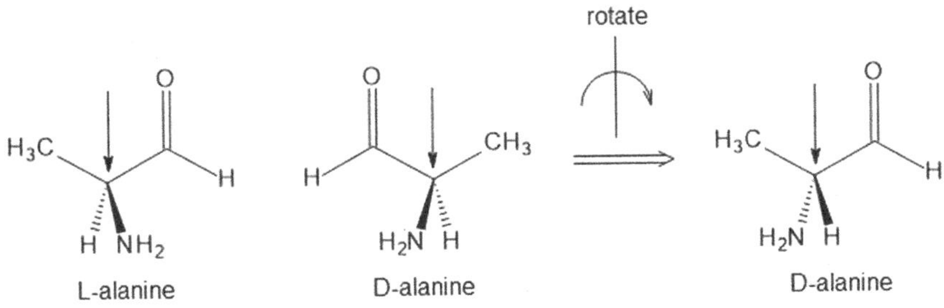

**Figure 7.7.** Mirror image relationship of alanines.

identical in their numbers of atoms of each element and in their bonding, but differ in how their atoms are arranged in space. We use the term *enantiomers* to describe two objects that are identical in all respects except not being superimposable on their mirror image (i.e., each is chiral). Enantiomers always come in pairs—the left hand and the right hand, the left foot and the right foot, the left shoe and the right shoe. Enantiomers are a subset of stereoisomers.

Molecules, too, can be chiral. In fact, the majority of biologically important molecules are chiral. For example, twenty common amino acids are used to construct the proteins in our cells and the cells of plants and animals. Of these twenty, nineteen are chiral. The simplest chiral amino acid is alanine, shown in Figure 7.7. The naturally occurring enantiomer is L-alanine, and its mirror image is D-alanine. (We'll get to what the D and L indicate shortly.) If you imagine a mirror perpendicular to the page sitting between the two molecules in Figure 7.7, you should see how one molecule could be reflected into the other. If we take D-alanine and rotate it 180° about a vertical line, we get the structure shown on the far right of the figure. This is still D-alanine; it's just flipped over. Now compare it with the structure of L-alanine on the left side. Both molecules have the C = O on the right side and the CH₃ group on the left side. All of that matches up perfectly. However, note the NH₂ group. In L-alanine, it is coming out of the page but in D-alanine it is positioned behind the page. These two molecules are not superimposable, and so each is chiral. Since they are the mirror image of each other, they are enantiomers.

The van't Hoff and Le Bel argument for the tetrahedral carbon atom essentially states that if a molecule has a carbon atom with four different groups bonded to it, the tetrahedral arrangement will force this molecule to be chiral. The top pair of molecules in Figure 7.8 has three different groups attached to a carbon atom. The two molecules are mirror images of each other. However, if you flip over the right molecule, you will note that it is identical to the molecule on the left.

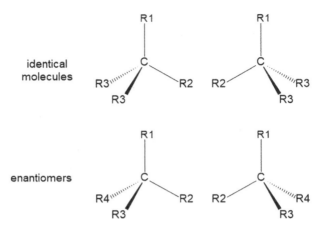

**Figure 7.8.** Identical molecules with three substituents and enantiomers with four substituents.

Now look at the bottom pair of molecules in Figure 7.8. Each molecule has four different groups attached to the carbon atom. These two molecules are mirror images of each other, and they are not superimposable. No rotations or flipping will take one molecule into the other. These two molecules are enantiomers. Almost every molecule that has a carbon with four different groups bonded to it will be chiral, and they will be one half of an enatiomeric pair. We call an atom with four different groups attached to it a *chiral center*. The arrows in Figure 7.7 point to the chiral centers in D- and L-alanine.

Enantiomers have identical physical properties, except one. The two enantiomers will have the same melting point, boiling point, solubility, energy, and so on. The one difference between enantiomers is how each interacts with polarized light. Plane polarized light is light that oscillates in a particular plane. This is the light that passes through polarized sunglasses. When plane polarized light interacts with a chiral molecule, that plane of oscillation will rotate. The sole difference between enantiomers is that the two molecules will rotate light in different directions—one clockwise, and one counterclockwise. Each enantiomer will rotate the light in the same amount but in opposite directions! That's it; otherwise they have the identical physical properties. We refer to this property as *optical activity*, where enantiomers will rotate polarized light in opposite directions.

The designations D- and L- are among a number of different systems we have for naming enantiomers. Some of these naming conventions (including D and L) are based on the physical structure of the molecule, and are effectively reminiscent of "right" and "left." The other naming conventions are based on whether the molecule rotates polarized light clockwise or counterclockwise.

This all may seem pretty esoteric, but there are practical applications. One of the critical decisions in winemaking, for example, involves deciding when to harvest the grapes. The winemaker will want to pick the grapes when they have a specific amount of sugar because the sugar will be converted to alcohol by yeast. Since the sugar in grapes is a chiral molecule, grape juice will rotate polarized light. The amount of rotation is linearly dependent on how much sugar is present. In the fall, you may see winemakers out in the grape vineyards looking through a tube (called a refractometer) filled with crushed grape juice, measuring the rotation of light to determine if the grapes have the correct amount of sugar and are ready to harvest.

What does chirality have to do with nucleophilic substitution reactions? Let's take a close look at the implications of a backside attack on a chiral molecule (Figure 7.9). The chiral substrate is represented generically with three different unreactive groups attached to carbon (R1, R2, and R3) and the leaving group. As the nucleophile approaches from the backside, it will start to push the three R groups to the right, along with helping to push out the leaving group. The R groups continue their move to the right until they have flipped over to the right side, with a new C–Nuc bond formed to the left. This process is akin to an umbrella being flipped upside down (or inverted) by a gust of wind. This motion was first described by the Russian-born (though now part of Latvia) physical chemist Paul Walden and is named *Walden inversion*.

Notice the structures of the reactant and product. If we (temporarily) consider the leaving group and nucleophile as being alike, then the reactant and product are mirror images of each other. The handedness of the reactant has changed (inverted) during the course of the reaction, switching from, say, the left-handed stereoisomer into the right-handed stereoisomer of the product. A substitution reaction that follows this backside attack will result in *stereoinversion*.

A frontside attack, which we suspect is not possible because the leaving group and nucleophile would get in each other's way will have a different

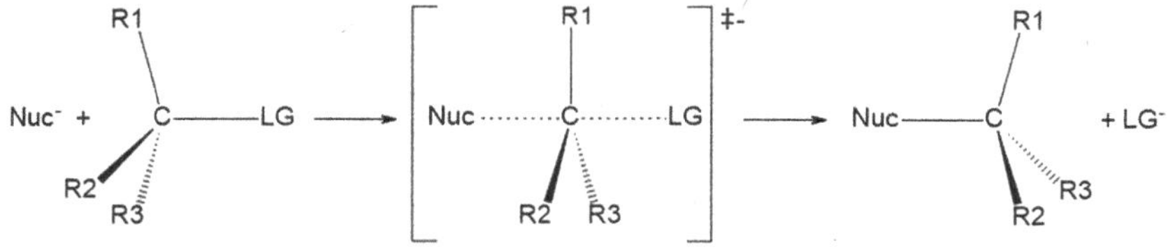

**Figure 7.9.** Walden inversion in an $S_N2$ reaction.

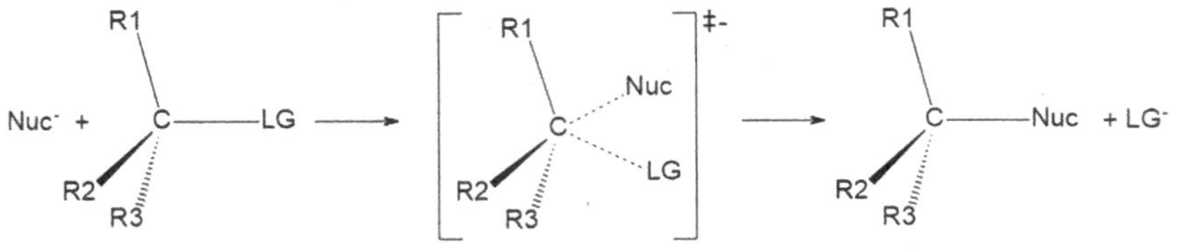

**Figure 7.10.** Stereoretention in a frontside S$_N$2 reaction.

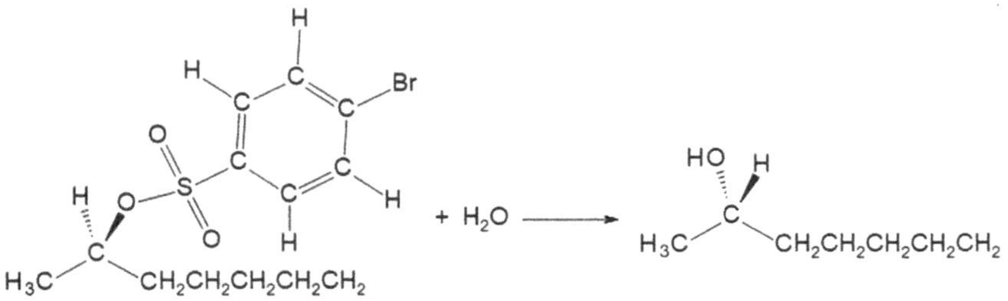

**Figure 7.11.** Experimental confirmation of backside attack in the S$_N$2 reaction.

stereo-outcome than a backside attack. This process is shown in Figure 7.10. The nucleophile coming in from the same side that the leaving group exits means that the stereocenter remains unchanged. If the reactant is, say, left-handed, the product will be left-handed too.

These two different outcomes imply that experiments are needed. Just what is the stereo-outcome of an S$_N$2 reaction? Stereoretention implicates a frontside attack, whereas stereoinversion implicates a backside attack. A clear study that established the stereochemistry for an S$_N$2 reaction at a chiral carbon is shown in Figure 7.11. In this reaction, water is the nucleophile and it has attacked from the back, opposite to the leaving group (that big group that reaches from oxygen to bromine), leading to complete inversion of the stereocenter. Backside attack is the mode for nucleophilic attack in the S$_N$2 reaction.

How about the S$_N$1 reaction? Does it have some stereochemical consequence? Substitution reactions involving chiral species under S$_N$1 conditions have been shown in many different experiments to lead to a racemic product. A racemate is a mixture of equal amounts of the left- and right-hand molecules. A racemic solution will not rotate polarized light; for every molecule that will rotate light clockwise, there is its enantiomer molecule that rotates light counterclockwise.

The two-step mechanism for the S$_N$1 nicely accounts for this loss of stereo-information with the production of a racemic product. This is outlined in

Figure 7.12. Suppose that we start with a single enantiomer of the reagent. The leaving group takes off, with a carbocation left behind. Now the carbocation has three pairs of bonding electrons about the central carbon. How can these three pairs be placed as far from each other as possible, while still being associated with the central carbon atom? The solution is to have them point into the corners of a flat equilateral triangle. This means that the carbocation is planar. This planar carbocation is represented in Figure 7.12, with the molecular plane coming in and out of the page.

In the next step of the $S_N1$ reaction, the nucleophile attacks the carbocation. The most direct, least hindered pathway is to approach the carbon perpendicular to the molecular plane. In Figure 7.12, these two perpendicular paths are drawn as the arrows from the left and right. Attack from these two paths leads to each of the enantiomeric products. These two paths are equally likely to occur, so each enantiomer is produced to the same amount. A racemic product is the result.

These sets of experiments provide strong evidence for the two competing reaction mechanisms. Their differing stereochemical outcomes—stereoinversion

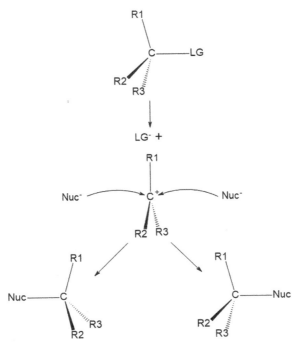

**Figure 7.12.** Racemic outcome of the $S_N1$ reaction.

for the $S_N2$ reaction and racemic products for the $S_N1$ reaction—definitively implicate two differing mechanisms, with specific constraints on the number of reaction steps and for molecular motion along the pathway. Delineation of these two mechanisms was among the first great achievements in physical organic chemistry.

# 8

# Interlude

## Philosophy of Science

I switch gears in this chapter and delve into the philosophy that guides the development of physical organic chemistry. It is this philosophy, this method for analyzing problems and constructing solutions, that is perhaps the most valuable attribute of physical organic chemistry applicable to disciplines beyond our narrow slice of the world.

The essential problem is this: organic chemistry is just so big. Chemical Abstracts Service (CAS) is a component of the American Chemical Society, the largest scientific society in the world. For well over a century, CAS has indexed the chemical literature, keeping track of essentially every molecule reported by every scientist. The CAS database has well over 163 million molecules, the majority of which are organic compounds. But this is just a small fraction of the estimated number of molecules that may exist. The GDB-17 database of chemical compounds contains 166 billion organic molecules, created by constructing all of the molecules that contain up to seventeen atoms of carbon, nitrogen, oxygen, sulfur, chlorine, or bromine. Even this is a small fraction of the expanse of chemicals, called *chemical space*. It is estimated that the total number of compounds is at least $10^{63}$ molecules. To give you some sense of just how huge that number is, the estimate of the number of atoms in the universe is between $10^{78}$ and $10^{82}$.

This is just part of the size problem. Organic chemists are interested in the transformation of molecules into new molecules. One might imagine taking a large list of molecules, mixing pairs of them, and then identifying when a reaction occurs and what is produced. Clearly, this database of reactions becomes extremely large as the number of reactants grows. Making sense of this amount of data is a distinct challenge.

How do scientists address problems? How do they gather and process information? Let's examine some of the key elements that govern how scientists do science.

*Thinking Like a Physical Organic Chemist.* Steven M. Bachrach, Oxford University Press. © Oxford University Press 2023.
DOI: 10.1093/oso/9780197640371.003.0008

## 8.1. Pattern Recognition

Humans are particularly capable of recognizing patterns. Douglas Hofstadter ably presented this concept in his work *Gödel, Escher and Bach*. He noted the amazing range of typefaces, yet we readily recognize the letters "a" and "A" in each of them (Figure 8.1).

The same principle applies to how we see the world. We recognize trees as such, even though there are over 60,000 different species of trees. We see commonalities in trees: they all have a trunk, branches, and leaves. In order to group them together, we can ignore the minor differences among them—for example, some have flowers, and not all of them lose their leaves in winter. In addition, we readily identify differences that separate trees from other plants such as bushes. Trees tend be taller and have a single main trunk, whereas bushes are typically less than about 10 ft tall and have multiple stems.

Countless examples can be given of how humans organize the world by identifying patterns. Through patterns, we recognize a cat as being different from a dog or we learn to identify the face of a friend in a crowd.

With the advancement of artificial intelligence over the past few decades, pattern recognition has taken on a more technical sense. Pattern recognition in this application is an outgrowth of statistics and signal processing. It is an attempt to use a variety of mathematical and computational techniques to assign some value (a number or a name or a category, etc.) to some new data based on relationships to some original set of data. For example, you may select a large set of pictures of cats, and from this training set, the computer program will build a set of traits that it associates with "cat." Then when you feed the program a new picture, it will scan the new image looking for the cat traits. If enough traits are identified, the program will assign the value "cat" to the image. If enough traits are not found, the program will assign the value "not cat" to the image.

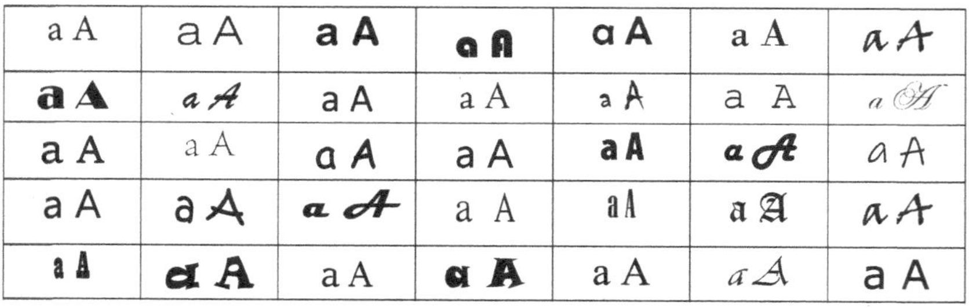

| a A | a A | a A | a A | a A | a A | a A |
|-----|-----|-----|-----|-----|-----|-----|
| a A | a A | a A | a A | a A | a A | a A |
| a A | a A | a A | a A | a A | a A | a A |
| a A | a A | a A | a A | a A | a A | a A |
| a A | a A | a A | a A | a A | a A | a A |

**Figure 8.1.** The letters "a" and "A" in different font faces.

Organic chemists identify patterns in many ways. We look for commonalities in structure: for example, the tetrahedral geometry about a carbon bonded to four different atoms, the planar geometry about a carbon bonded to three different atoms, and the linear geometry about a carbon bonded to two different atoms. The planarity of aromatic compounds (to be discussed in Chapter 16) and the linear arrangement of atoms in a hydrogen bond.

Organic chemists also look for similarities of reactions. For example, we note that a variety of different nucleophiles all behave in the same way when attacking a carbon with a (partial) positive charge. Or as we'll see in Chapter 17, we study the myriad different reactions involving molecules with double bonds where two molecules combine to form a ring or when a single molecule rearranges into a ring.

The power of pattern recognition for organic chemists is its ability to enable the grouping of related molecules or related reactions and to find relationships between the groups. This leads us directly to our next topic: using patterns to assign categories and hierarchies.

## 8.2.  Categories and Hierarchies

Fasteners are devices that hold objects together. Consider the fasteners used in construction: nails and bolts and screws. There are roofing nails and siding nails and dry wall nails and finishing nails. We have j-bolts and u-bolts and eye bolts and shoulder bolts and lag bolts. There are wood screws and machine screws and thread cutting screws and sheet metal screws and dry wall screws and socket screws and set screws. These can come with flat heads and round heads and hex heads and pan heads and oval heads and button heads. They have slotted and Phillips and hex and torx and square drives.

We can create a hierarchy of fasteners, subdividing them based on common features and differences between types. At the top, we group all the fasteners into one giant pile. As we move downward in the hierarchy, we separate by category, getting more and more distinctive with each level.

Scientists create hierarchies to help define similarities and differences in order to show relationships in recognizable ways. Perhaps the most famous scientific hierarchy is the taxonomy of living organisms shown in Figure 8.2. At the top, only extraordinary differences separate organisms. Mammals and plants and insects are in the same domain (*Eukarya*), but bacteria are in a different one (*Bacteria*). As one moves down the hierarchy, more subtle differences start to separate the organisms. A fly and a human are in the same kingdom (*Animalia*) but different phylum (*Anthropoda* for the fly and

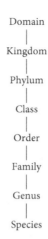

Domain
|
Kingdom
|
Phylum
|
Class
|
Order
|
Family
|
Genus
|
Species

**Figure 8.2.** Taxonomy of living organisms.

*Chordata* for the human). A human and a chimpanzee are in the same family (*Hominidae*), but the chimpanzee is in genus *Pan,* while humans are in genus *Homo.* At the very bottom are organisms that are quite closely related, such as humans (*Homo sapiens*) and Neanderthals (*Homo neanderthalensis*)

Chemists create hierarchies in many places. The first hierarchy often taught in college chemistry defines matter. At the top are mixtures, which are made up of pure substances also known as compounds. Compounds are made up of molecules; molecules are made up of elements; elements are composed of atoms; and atoms contain three subatomic particles: neutrons, protons, and electrons.

The essential element for identifying categories of organic compounds is the *functional group,* which is an atom or a collection of atoms whose physical and chemical properties differ little depending on the rest of the molecular environment. The molecule ethanol **1** is the intoxicating ingredient in alcoholic beverages such as beer and wine. It contains a hydroxy group, an oxygen bonded to a hydrogen atom. Isopropanol **2**, a disinfectant, also contains an OH group. Ethylene glycol **3** is used in antifreeze. Glucose **4** is the most common sugar.

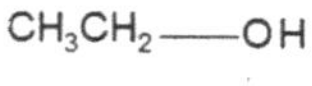

$CH_3CH_2$—OH

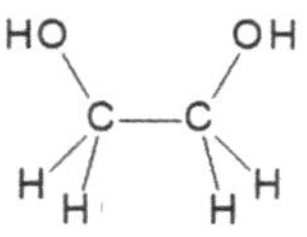

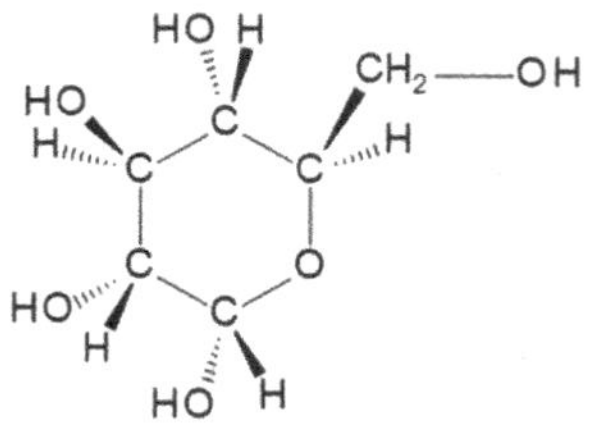

**1**

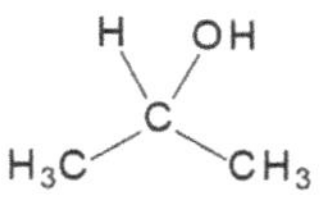

All of these molecules have at least one OH group, and we group them to-gether as *alcohols*. They all have many commonalities in terms of both their physical properties, like solubility in water and relatively high melting points, and their chemical properties, such as their ability to form ethers and alde-hyde and carboxylic acids.

This last relationship is most important to organic chemists. Identification of different functional groups and the methods for chemically converting one functional group into another provides the core pedagogy for how organic chemistry is taught today. All textbooks pretty much follow this same pattern of introducing one functional group chapter by chapter and then discussing how the previously mentioned groups might be converted into the new one and how this new functional group might be converted into other functional groups.

Instead of thinking about and remembering a whole series of specific reactions, such as how to convert ethanol to acetaldehyde, *n*-propanol into propanal, and *n*-butanol into butanal (see Figure 8.3), we can group them all together. The three reactions shown in this figure utilize different oxidizing reagents (PCC, DMP, or TPAP) to increase the oxygen content in the mol-ecule. (While the oxygen content may not seem to have changed in these reactions since all the reactants and products have only one oxygen atom, one can consider that there is twice the oxygen content in the product due to the C–O double bond.) In many situations, the choice of which oxidizing reagent to use is arbitrary, and many alternatives are available. The key recognition here is that in all of these examples, and many more, an alcohol is oxidized to an aldehyde. This can be generalized as the bottom reaction of Figure 8.3. The symbol [O] represents the use of an appropriate oxidizing agent. Since there are thousands of different alcohols, this grouping certainly makes it easier to manage!

Close examination of the reactions presented in Figure 8.3 should give some pause, especially if you remember what is perhaps the most important lesson of introductory chemistry. These reactions are not balanced! A major theme of introductory (or general) chemistry is that mass cannot be cre-ated or destroyed. Applied to chemistry, this theme is extended to conserva-tion of atoms; namely, in a chemical reaction, the atoms cannot be created

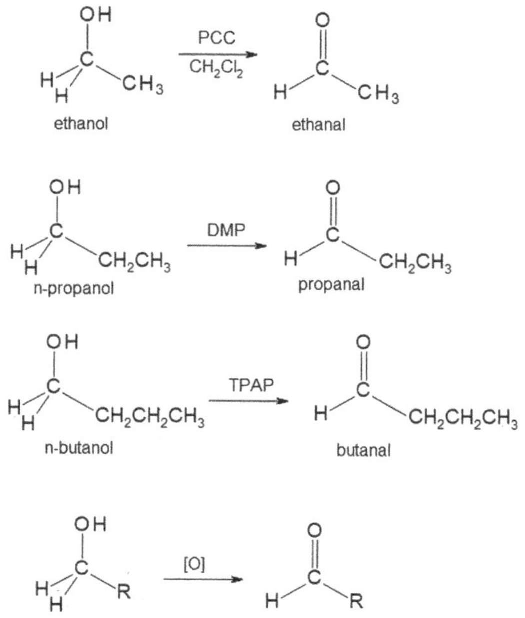

**Figure 8.3.** Oxidations of alcohols to aldehydes.

or destroyed. For example, if we start out with reactants that have six carbon atoms, a dozen hydrogen atoms, and four oxygen atoms, the products must have the exact same number of atoms of each element—six carbon, twelve hydrogen, and four oxygen atoms. We refer to this as a *balanced reaction*. Organic chemists, however, get a bit sloppy. We really only care about what happens to the molecules that contain carbon. We will include all of the reactants, but if a product is an inorganic molecule, such as NaCl or water or $H_2$, we will typically omit it from the written reaction. It's not that these reactions violate the rules; it's just that we want to focus our attention on the important aspects of the reaction.

Functional group transformations like the conversion of alcohols to aldehydes given in Figure 8.3 inspire us to think of other relationships. For example, might it be possible to perform the reverse transformation: converting aldehydes to alcohols? In fact, this is possible, using a reducing agent, such as $LiAH_4$ or $NaBH_4$. We might write this transformation as shown in Figure 8.4, with the term [H] representing an appropriate reducing agent. Looking at the bond changes in the reduction reaction, we find that the C–O double bond

Figure 8.4 Reduction of aldehydes to alcohols.

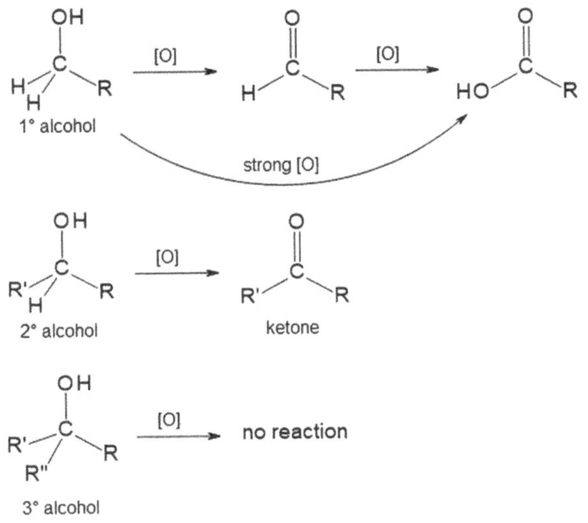

Figure 8.5. Oxidation of primary, secondary and tertiary alcohols.

is lost and that a C–H and O–H bond are formed. It's as if $H_2$ has been added across that double bond, and that's where the [H] term comes from.

Hierarchies within functional groups are also valuable. We can divide the alcohols into three groups based on their degrees of substitution: primary, secondary, and tertiary alcohols. These groups can help us further relate oxidation reactions. As we have seen, primary alcohols can be oxidized to aldehydes. Aldehydes can be oxidized to carboxylic acids, so with strong oxidizing agents, we can directly convert primary alcohols to carboxylic acids. Secondary alcohols can be oxidized to ketones. Careful examination of these reactions reveals that these oxidation reactions remove hydrogen from the substituted carbon. Tertiary alcohols have no hydrogen on the substituted carbon, and so they are not susceptible to oxidation reactions. All of these reactions are summarized in Figure 8.5.

## 8.3. Models

Scientists rely on experiments to obtain information. Experiments might be so simple that they can be performed in an ordinary household. Recall, for instance, physicist and Nobel Laureate Richard Feynman demonstrating that rubber o-rings lose their flexibility at cold temperatures by dunking one in a glass of ice water, thereby proffering an explanation for the cause of the Space Shuttle *Challenger* explosion. On the other extreme, experiments might require construction of huge complexes that employ a thousand scientists and technicians, such as the Laser Interferometer Gravitational-Wave Observatory (LIGO) experiment that first detected gravitational waves.

The best experiments are designed to unambiguously provide data to answer a simple question, often using a control experiment to compare with the null result. More often, experiments provide data that needs interpretation, to be followed up with duplicate tests or additional experiments.

Careful experimental design is the calling card of scientists. We try to anticipate how the experiment is supposed to work, and then we troubleshoot in advance a number of potential ways the experiment can go awry. We want to control as many aspects of the experiment as possible, so that optimally a single idea, a single *variable*, is being tested. We might set up an experiment such that all the conditions might retard the hoped for outcome in order to put the idea under a stress test. Absolutely essential to experimental design is that it can be replicated, not just in the same lab, but by another scientist half a world away. I highlight these features of well-designed experiments throughout this book.

Conclusions can be drawn based on the results of many experiments conducted by many different scientists in many different locations. More importantly, a model can be constructed that explains the results in the context of what has previously been done. Traditionally, this process is referred to as constructing a *theory* that provides understanding and predictions. I choose to use the term *model* here instead. As for the term *theory*, scientists well understand it to indicate a set of concepts, rules, and equations that explain a major portion of a discipline. The ramifications of a theory have been extremely well tested, and the results have been duplicated many times over. A theory is as close to truth as a scientist gets. Think of Einstein's theory of special relativity or general relativity, or the theory of quantum mechanics, or the theory of plate tectonics. As much as we hold these theories in high regard as part of the established scientific canon, exactly the same gravitas is accorded to the theory of evolution. And there's the bugaboo.

No nonscientist questions the theory of relativity because Einstein is popularly considered the smartest person ever. And quantum mechanics is so weird and cool and doesn't seem to apply to the real world anyway, so we'll let that slide. As for plate tectonics, if anyone thinks about it at all, the picture of Africa and South America nestled together just makes intuitive sense.

But the theory of evolution confronts some of the foundations of our civilizations and societies. It conflicts with the take of many religions on human origins and requires us to question our relationship with all living things. The nonscientist hears that word *theory* and confuses it with its common meaning: a theory is just "my belief" or "my conjecture" or "my explanation." This meaning of the word *theory* carries none of the weight of countless repeated experiments that demonstrate the utility and validity of a scientific *theory*. In fact, it completely ignores that crucial detail. If I can have a theory as to why my favorite football team lost this past Sunday, then the theory of evolution is equally arbitrary and subject to dispute.

The term *model* is perhaps less loaded with ambiguity. We all recognize that a model approximates something. A model car is a replica of a car; it captures some of the features of the original car—its styling, the relative size of the tires, the shape of the exhaust system—but no one expects it to deliver all that the original car can do. No one expects to hop into a 1/36 model of a Ford Mustang and drive down the street. The model ship of the *Titanic* displayed in a museum is not expected to be seaworthy and perhaps be able to avoid an iceberg.

A scientific model is also an approximation, capturing some component of the real world around us. Newtonian mechanics is a model of how objects move within the assumptions of the classical world. It works extremely well to understand how the cue ball collides with the 8-ball, which then rolls into the side pocket. Classical mechanics is fine for calculating the trajectory of projectiles, allowing the artillery unit to deliver ordinance on target.

Mercury, like all planets, orbits the sun in an elliptical path. This is well described by classical mechanics. The perihelion of Mercury, its closest approach to the sun, does not occur in the same pass with each orbit. Rather, it advances in a circular-like pattern. This is called the precession of the perihelion, and precession is also predicted by classical mechanics. The motions of all of the planets other than Mercury are described quite ably by classical mechanics. The precession of Mercury, however, is about 1% faster than that predicted by classical mechanics. This is a failure of classical mechanics, one that is corrected by Einstein's theory of general relativity. This failure defines one of the boundaries of classical mechanics—motion in large gravitational fields is outside the realm of classical mechanics. Classical mechanics also fails to properly describe when objects move extremely fast; here we need the corrections provided by special relativity. It fails to also properly account for the motion of very tiny objects, like subatomic particles; quantum mechanics is required in these situations.

For me, the term *model* seems apt in describing classical mechanics. Classical (Newtonian) mechanics works very well for many situations but breaks down in known and predictable ways. We don't expect a "model" to be perfect and error-free. We recognize it as an approximation, and science moves by ever improving upon these approximations. The best we scientists can do is to devise a really good model, but perfect understanding of the world appears to be beyond our abilities but remains our quest.

The term *model* is in some use in science already. The nondescript moniker *Standard Model* is in fact one of the most robust theories within physics. The Standard Model provides the theoretical framework that unifies the weak and the strong forces (both of which are used to describe the forces that hold a nucleus together) and the electromagnetic force. It provides an understanding of subatomic particles, including quarks. The Standard Model was built on significant, highly reproduced experiments, and its greatest triumphs were the predictions of the existence of new particles that had not yet been observed but were detected subsequently through monumental experimental effort. The Standard Model is the outcome of myriad scientists whose individual work built up this masterful understanding. For all its tremendous successes—the prediction and subsequent discovery of the $W$ and $Z$ particles and the Higgs boson—the Standard Model has many known limitations. It fails to incorporate gravitation, the last remaining force, and it offers no explanation for the expanding universe or for the possibility of dark matter and a dark force.

The Standard Model leaves many philosophical questions unanswered. Why *is* there a Standard Model? Why do particles have their specific mass? What gives rise to the asymmetry of the universe such that it contains more matter than antimatter? All of these limitations and unanswered questions do not signal that the Standard Model is wrong or inadequate. Rather, we know we have more science to do! This is how science works: we conduct lots of experiments, develop a model that makes predictions, test to see if those predictions are confirmed or not, and continue the cycle. The outcome is ever more comprehensive models and an ever-growing understanding of the world. Science is an ongoing process.

Let's take a walk through that model building and testing process. A nice example of a scientific model is the kinetic theory of gases that culminates in the ideal gas law. The experimental studies of gases constitutes some of the earliest works of what might be considered modern physical science. There are three major experimental laws of gases, all of which were discovered centuries ago.

First, in 1662, Robert Boyle and his assistant Robert Hooke identified the inverse relationship of pressure and volume. A simple way to imagine this relationship is to hold a balloon in your hands. If you squeeze the balloon, you are applying more pressure to the gas inside, and the response is for the volume of the balloon, the volume of the gas inside, to decrease. Ease up on the pressure and the balloon expands back. This relationship is known as Boyle's Law.

Second, in the 1780s, Jacques Charles discovered what is now known as Charles's Law, but he did not publish his work. (Apparently, he was not concerned about tenure.) In the early 1800s, two giants of chemistry, John Dalton and Joseph Louis Gay-Lussac, published confirming reports of this law of gases. Returning to our balloon, when placed in an ice bath, it shrinks. Remove it from the ice bath, warm it up with a flame, and the balloon expands until the rubber breaks and the balloon explodes. This experiment demonstrates the direct relationship between temperature and volume: with increasing temperature, the volume of a gas will increase.

The third law was proposed by Amadeo Avogadro in 1811. Subsequent experimental work by Charles Frederic Gerhardt and Auguste Laurent confirmed the hypothesis, which is now known as Avogadro's Law. Let's return one last time to our balloon. Open it up and blow into it, adding more gas to the balloon. What happens? The balloon expands in volume, demonstrating the proportional relationship between the amount of gas and its volume. (I realize this experiment seems so trivial; why would someone merit getting his name on that law? To be fair to Avogadro, his hypothesis was much more profound: an equal number of molecules of any compounds would have identical volumes. But for our development here, the simpler experiment—and straightforward outcome of Avogadro's hypothesis—suffices.)

These three gas laws were later combined by Emile Clapeyron to form the *ideal gas law*, which today is the subject of many homework problems and test questions in every introductory chemistry class. The ideal gas law is quite simple: $PV = nRT$. Multiply the pressure ($P$) and the volume ($V$) of a gas, and it will equal the value of the product of the number of molecules ($n$), the gas constant ($R$, an experimentally derived number), and the temperature ($T$).

The ideal gas law can also be obtained using statistical mechanics. Starting with classical physics, James Clerk Maxwell considered that the large number of molecules in a gas held within a jar or balloon would allow for the application of statistics to this ensemble. This led him to derive a distribution of the velocities of the collection of molecules in a gas. Ludwig Boltzmann later

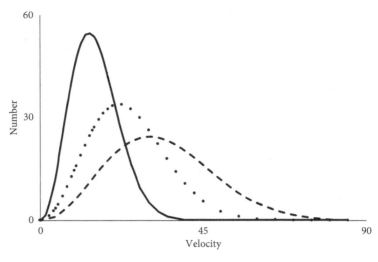

**Figure 8.6.** Boltzmann-Maxwell distribution of gas molecules. The solid curve is at low temperature, the dotted curve is at medium temperature and dashed curve is at high temperature.

reformulated these notions to create the Boltzmann–Maxwell distribution. From this distribution, the ideal gas law can be derived.

The Boltzmann–Maxwell distribution is displayed in Figure 8.6. The $x$-axis marks the increasing velocities of the gas molecules, and the $y$-axis indicates the number of molecules with that particular velocity. There are very few molecules at rest, and the number of molecules with a particular velocity climbs until it reaches a maximum, after which the number of molecules with further increasing velocity starts to diminish. There are very few molecules with extremely large velocities. Perhaps more interesting are the changes to the distribution with increasing temperature. As the temperature of the gas rises, there are more molecules moving faster; in other words, the distribution curve shifts to the right, to higher velocities. This change in distribution, namely, more molecules moving faster as the temperature rises, will have important consequences that we will revisit many times in later chapters.

So why is this called the *ideal* gas law? What's ideal about it? Here is where the notion of a model, a simplification, an *idealization* of the real world, comes to play. Let's simplify the world and consider gases to be made up of small particles (molecules) that fly around, most of the time widely separated from each other. The vast majority of the volume of the gas is empty. Occasionally, the gas molecules will run into the walls of a container, but the molecules will simply bounce right off. The molecules may also collide with other molecules, but again the molecules will bounce off of each other. The molecules won't stick to each other or to the walls; these are *elastic collisions*. Furthermore, the molecules don't react with each other. Under these conditions, the ideal gas law works perfectly.

Many gases do behave "ideally." At high temperatures and low pressures, the ideal gas law will apply for almost all substances. It's under the opposite conditions that problems arise, that nonideal behavior is seen.

Let's think about what happens to a gas at low temperatures. If we go back to Figure 8.6, we see that as temperatures decline, the velocities of the gas molecules shift to lower values. As the molecules move more slowly, they have less kinetic energy. Neutral molecules all have a slight attraction to each other, and with less kinetic energy available, at low temperatures the molecules will start to stick together.

Consider steam, water, in the gas phase. As it cools, it begins to condense, forming liquid water. This phase change happens because the molecules do not have sufficient kinetic energy to overcome the inherent attraction between water molecules. Certainly, one should not expect any equation that describes a gas to apply to a liquid. Therefore, gases become nonideal, more liquid-like, at lower temperatures.

The same logic applies at high pressures. Recall Boyle's Law, which states that as pressure increases on a gas, its volume will decrease. As the volume decreases, the molecules are forced to be close together, and again the inherent attraction between molecules will mean that the gas will start to condense, forming a liquid.

Nonideal gas behavior really means that the gas is exhibiting some behaviors that mimic a liquid or even a solid. Our ideal gas law is a model: it applies to only specific situations, but these specific situations are commonly found. The model is very useful, but one always needs to recognize its limitations.

Why might we continue to use a model that we know is wrong, where we have a better, more robust model? Sometimes, it's a convenience. If we are looking at the motion of billiard balls on a pool table, quantum mechanics will yield the same result as classical mechanics, but the quantum mechanical solution will be significantly more difficult to compute than with classical mechanics! In some situations, the more advanced model is simply not needed; the correction it applies is not applicable in the given circumstance. For example, the conservation of mass is superseded by the conservation of mass-energy, which incorporates relativity. If we are examining a chemical reaction, then conservation of mass will be the exact answer; there is no mass-energy conversion going on. However, if we examine a nuclear reaction, than we must apply the conservation of mass-energy.

Models allow us to capture the essence of a problem, ignore some parts that we hope will not be missed, and thereby provide some understanding, guidance, and predictability. Models compress a complicated world into a manageable idealized realm.

A powerful model is one that is easy to understand, broadly applicable, makes predictions that can be tested, and has well-known boundaries. The successful models hit all of these marks. The failed models miss on one or more of these aims, though these failures often lead to a revised model that is much improved.

What is meant by "well-known boundaries"? It is understanding the conditions under which the model will apply, when to start to be concerned that the model might break down, and under what conditions the model will truly fail. Here's an example. Remember your high school geometry class and the major lesson on triangles. The sum of the three angles in every triangle is 180°, no more and no less.

Let's see how this works here on earth, which is more or less a sphere. Imagine yourself at the North Pole. Now walk south along the Prime Meridian, through Greenwich, and all the way to the equator. Make a right turn and walk across the Atlantic Ocean (OK—walking on water is part of this imaginary exercise), across South America onto the Pacific Ocean all the way to the 90° west longitude line. Make another right turn, heading north across the United States and Canada, back to the North Pole. This is the highlighted path on the sphere in Figure 8.7. The path is a triangle; you've

**Figure 8.7.** Triangle on a sphere.

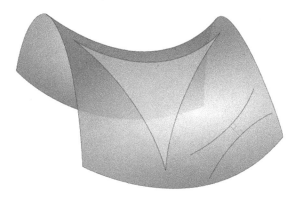

**Figure 8.8.** Triangle on a saddle.

walked on three straight lines: south along the 0° meridian; west along the equator; and north along the 90° W meridian. Now what are the angles in this triangle? At your first turn, it's 90°, and so it is at your second turn. Back at the North Pole, it's the intersection of the 0° and 90° W meridian, so that angle is 90° as well. The sum of the angles in this triangle on a sphere is 270°! When we work with triangles on a sphere, we learn that the sum of their three angles is always greater than 180°. Mathematicians refer to this geometry as *spherical geometry*.

Although it is more difficult to imagine, what is the sum of the angles of a triangle on a saddle? Although the sides of the triangle may seem bent, if you were walking on the saddle surface (Figure 8.8), these are the shortest paths connecting the three vertices. They are the "straight lines" on a saddle. The sum of the angles of any triangle on a saddle is less than 180°. This is the world of *hyperbolic geometry*.

What we learned in our first geometry class was *Euclidean geometry*, the geometry of a flat (planar) surface. On a flat surface, the angles of a triangle always sum to 180°. Any deviation from a flat surface leads to triangles that have angle sums that are either greater or less than 180°. This is the boundary, or limit, of the validity of Euclidean geometry. Exploration of this boundary led to some amazing mathematics, for example, dispelling the notion that the fifth postulate of Euclidean geometry (the famous parallel postulate) could be proven from the other four postulates.

In all disciplines of science, this constant probing at the boundaries of models has repeatedly improved them, providing understanding of Nature. This leads to the next important component of the scientific method: testing.

## 8.4. Testing

The twentieth-century philosopher of science Karl Popper espoused *critical rationalism* as the basis of science. He suggested that a scientific concept could never be proven correct and that it could only be consistent with known experimental results. Rather, the key element is that a scientific concept must be *falsifiable*, that some experiment could be undertaken that could end in a negative result. Science must be subject to ongoing tests. A single negative result will be the concept's undoing; we are compelled to identify and conduct experiments that could manifest in a negative result. The acerbic Nobel Laureate of Physics Wolfgang Pauli is noted for his most cutting putdown, "not even wrong," to characterize a theory or model that can't even be tested, truly capturing this requirement of falsifiability.

A terrific example of this continuous testing process is the muon g-2 experiment reported in the spring of 2021. The muon is a subatomic particle very much like an electron, except that it is about 200 times heavier and lives for a very short time, about 2 microseconds. Just like an electron, muons have spin, which means they behave like tiny magnets. When placed in a magnetic field, electrons will wobble like a spinning top (what we call precession) and so will muons. The Standard Model can be used to calculate the rate of this wobble. In 2001, experiments at Brookhaven National Laboratory indicated a disagreement in the rate of the muon wobble predicted by the Standard Model. This experiment was repeated with greater precision in 2021 at Fermilab, and these results confirm the Brookhaven results. The disagreement between the experimental and theoretical rates derived from the Standard Model points to some error in the model. It may be that there are other particles or other forces in nature or some other new physics to be discovered. Perhaps there is some error in the experiments themselves, and continuing experiments are underway. We should obtain clarification regarding the accuracy of the experiment in another year or two. The response within the physicist community is nothing short of giddiness, as this result points us toward new discoveries and improved models. It's the way science gets done!

All the physical organic chemistry models I discuss in this book have been subjected to myriad tests. These tests probe the strength of the model, especially at the boundaries, allowing chemists to refine the models or develop new ones.

The process of collecting data, proposing hypotheses, testing, creating a model, and then conducting rigorous follow-up experimentation defines the scientific method. Physical organic chemists are especially talented in designing clever experiments that probe the edges of the models. It is this type of thinking that I want to highlight as it can be broadly applied.

## 8.5. Beauty and Symmetry

Nonscientists may be surprised to learn how often scientists refer to beauty and symmetry in guiding their thinking. The physicist and Nobel Prize winner Paul Dirac, who united quantum mechanics and special relativity through what became known as the Dirac Equation, was asked how he found this equation. His response: "I found it beautiful." Physics Nobelist Richard Feynman has this to say about the Euler identity: "We summarize with this, the most remarkable formula in mathematics: $e^{i\theta} = \cos\theta + i\sin\theta$. This is our jewel." Periodically, *Science* magazine polls scientists for what is the most beautiful equation; nominees invariably include Einstein's famous equation $E = mc^2$, or Maxwell's equations that define electromagnetic fields, Euler's identity, and many others.

Leave it to scientists to use scientific methods to assess the aesthetic response to mathematics. magnetic resonance imaging (MRI) is a noninvasive tool for examining soft tissue and bone within a live being. It you have recently suffered an injury to a joint or back, your doctor probably had you get an MRI. You sat inside a big tube–really a large magnet—and radio waves were blasted into your body, revealing your interior without necessitating exploratory surgery.

A recent extension of this examination, called functional MRI (fMRI), uses the technique to explore brain activity. In a typical experiment, a person is inside the tube of the fMRI and looks at pictures or listens to sounds. The fMRI monitors brain activity, and the scientists seek to learn which regions of the brain "light up," showing a response, to that particular stimulus. For example, neurologists have identified certain regions of the brain that respond, that show activity, when a person sees a beautiful picture or hears a lovely musical passage.

The British interdisciplinary team of two neurologists, Semir Zeki and John Paul Romaya Dionigi, along with the physicist M. T. Benincasa and the mathematician Michael F. Atiyah, supposed that mathematicians would have the same brain response to a beautiful equation as when a musician hears a passage by Mozart or when an artist looks at a painting by Van Gogh. Their 2014 investigation began with a list of sixty equations that sixteen mathematicians had ranked as beautiful, neutral, or ugly. These same mathematicians were then placed in the fMRI, and their brain activity was monitored as they were shown these sixty equations. The beautiful equations caused the same location in the mathematician's brain to respond as when artists or musicians responded to beautiful works of art or music. At least on a neurological plane, beauty is beauty, whether it's a poem, a sculpture, a symphony, or a mathematical equation.

What makes an equation beautiful or ugly? In exploring this question, we find ourselves on treacherous terrain. What makes a sunset sublime; a poem lyrical; a painting beautiful; a sonata magnificent? Let's take a look at what those sixteen mathematicians selected as the most beautiful and the ugliest equation.

The equation judged the most beautiful is a slight modification of the Euler identity: $1 + e^{i\pi} = 0$. The authors of the study describe this equation as follows:

> Euler's identity links five fundamental mathematical constants with three basic arithmetic operations, each occurring once.

It's not just that there are five constants in this equation; the important point is that all of the most important, the most widely known, the most versatile of the constants are represented. The numbers (constants) zero and one are essential for counting. There's Euler's number $e$, the base of the natural logarithm, which is found in wide applications such as calculating compound interest and computing essential integrals. The equation includes the square root of $-1$, $i$, which leads to the imaginary numbers. And last, maybe the most famous mathematical constant is $\pi$, the ratio of the circumference of a circle to its diameter. All of these in one simple equation that also happens to involve the addition, multiplication, and power operations. So simple, all those magical numbers in one short statement. How is that *not* beautiful?

What unfortunate example is the ugliest equation? Keep in mind that this equation is the ugliest of the sixty and *not* the ugliest equation ever. I mean the equation $6 - 4 = 3$ is really ugly; it's false! The mathematicians judged this formula, presented by Srinivasa Ramanujan, the famous Indian mathematician, as being the ugliest:

$$\frac{1}{\pi} = \frac{2\sqrt{2}}{9801} \sum_{k=0}^{\infty} \frac{(4k)!(1103 + 26390k)}{(k!)^4 \, 396^{4k}}$$

This equation is not ugly because it is an infinite series. Many other equations on the list of sixty involve infinite series. One of the other sixty equations also provides a means for computing $\pi$:

$$\frac{\pi}{4} = 1 - \frac{1}{3} + \frac{1}{5} - \frac{1}{7} + \frac{1}{9} \cdots$$

and it's not judged to be ugly. This expression for π has a symmetry and pattern that are readily apparent. But Ramanujan's formula, is judged to be ugly because the numbers seem so random, so arbitrary—why 9801, why 26390? What's the reasoning behind those factorials, and a factorial to the fourth power? It's also computationally messy, not readily computed, and certainly not readily evaluated by using paper and pencil. My guess is that it's ugly because it just doesn't seem to make sense and doesn't seem to connect to other equations in a natural way.

Ramanujan's ugly formula happens to be quite efficient in computing π. Each added term picks up about eight additional decimal places. This formula inspired some of the work done by mathematicians in the last thirty years to dramatically improve the ability to compute π to an extraordinary number of digits. So, perhaps beauty here is more than skin deep.

Chemists seem to find beauty in symmetry. The platonic solids have inspired humankind for millennia; for chemists, these objects inspired imaginative efforts that led to the synthesis of a substituted tetrahedrane, cubane, and dodecahedrane (Figure 8.9). The 1996 Nobel Prize in Chemistry was awarded to Robert Curl, Richard Smalley, and Harold Kroto for their discovery of fullerene (Figure 8.10). Fullerene has sixty carbon atoms, connected in the shape of a soccer ball. This is not a Platonic solid but possesses considerable symmetry. The attractiveness and attention-generation of these molecules is certainly tied to their high degree of symmetry.

Symmetry is a key component of stereoisomers, particularly the relationship of enantiomers discussed in Chapter 7. Significant efforts have been made over the past two decades to develop techniques for synthesizing one enantiomer over the other, which requires breaking symmetry.

In Chapter 17, I present the story of pericyclic reactions, one of the crowning achievements of physical organic chemistry. This work is rooted in symmetry, particularly the role symmetry plays in quantum mechanics.

The point of this discussion is that notions of aesthetics—beauty, symmetry, simplicity, elegance—are factors at play when scientists develop and assess theories. Scientists are not immune to emotional responses. A theory that concisely captures a problem with a simple explanation or a well-formed equation will naturally engender a positive response. The mathematician and philosopher Bertrand Russell noted that "the true spirit of delight, the exaltation, the sense of being more than Man, which is the touchstone of the highest excellence, is to be found in mathematics as surely as poetry."

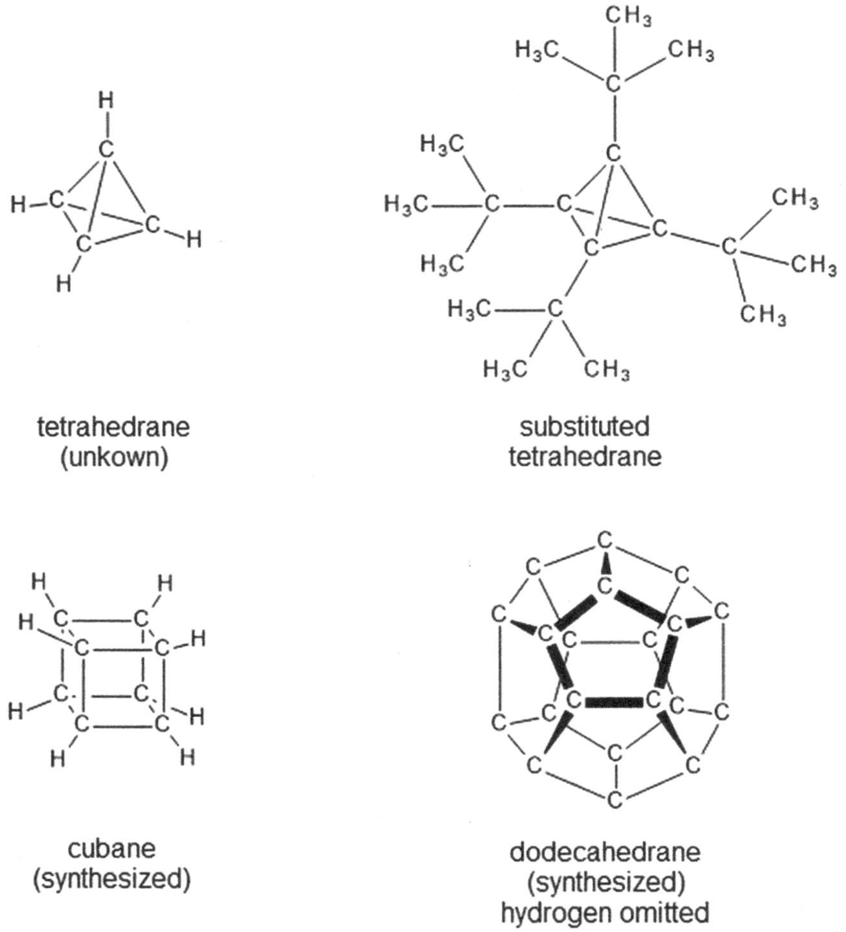

tetrahedrane
(unkown)

substituted
tetrahedrane

cubane
(synthesized)

dodecahedrane
(synthesized)
hydrogen omitted

**Figure 8.9.** Chemical compounds mimicking the Platonic solids.

## 8.6. Occam's Razor

A set of experimental results can often be interpreted or explained in many different ways. That's actually one of the great features of science; it's a great debating society. Competing ideas are placed in the marketplace to be judged by scientists. How well does the proposal agree with the known experimental results? What experiments does the proposal suggest be undertaken next— and then how do the results of those next experiments match up with the theory? There are always more questions to ask, more data to gather, and more refinements of the hypothesis.

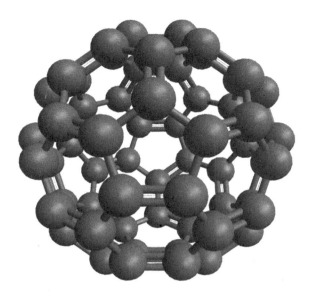

**Figure 8.10.** 3-D representation of fullerene. Each ball indicates a carbon atom.

One of the guiding principles for assessing competing hypotheses is Occam's Razor, which states essentially that given multiple explanations, the one that has the fewer assumptions should be chosen. Put another way, one should choose the explanation that is simpler. Occam's Razor is not absolute. Sometimes complications are needed as the simpler explanation just misses out on too much of the real world. Setting the value of π equal to 3 would certainly make computations easier, but the error is simply too large for you to live with; certainly, you would not want to drive across an arch bridge designed by a civil engineering using a value of three for π. Occam's Razor is more of "good practice," a way to guide the development of explanations toward solutions that both work and are understandable.

An extension of Occam's Razor is a famous quote attributed to Albert Einstein, "Everything should be made as simple as possible, but no simpler." As with many famous lines, this one apparently was never spoken or written by Einstein. It is a paraphrase, in a much simpler (!) way, of the likely origin of the idea,

It can scarcely be denied that the supreme goal of all theory is to make the irreducible basic elements as simple and as few as possible without having to surrender the adequate representation of a single datum of experience.

Einstein made this statement in the Herbert Spencer Lecture he delivered in 1933 titled "On the Method of Theoretical Physics."

A good theory, a good model, is simple, readily understood, broadly applicable, and testable. But it should not be too simple; do not sacrifice agreement with experiment for a theory or equation that just appeals to one's notion of simplicity.

## 8.7. Not a Proof

Let's turn once again to Karl Popper as we begin the last topic of this chapter. Popper wrote: "In the empirical sciences, which alone can furnish us with information about the world we live in, proofs do not occur, if we mean by 'proof' an argument which establishes once and for ever the truth of a theory." This may strike the layperson as audacious, for isn't science all about "proof," about establishing what is fact and what is fiction, and the laws that govern the operations of the universe?

Proof implies finality, the absolute assurance that the proposed explanation is complete and true. Science never reaches this point. Experiments provide facts and reproducible data, which can be confirmed through repetition; hypotheses provide a potential explanation of some facts; and theories provide explanations that cover a broad range of facts, consistent with all of the evidence we have gathered so far. Nonetheless, there is always that next experiment out there, with results that might conflict with the wisdom of the day. We can never be sure that new data is just around the corner that will require us to rethink our previous theories.

Science historians use the language of revolutions, speaking of how one theory overthrows another, thereby creating a dramatically new way of seeing the world. Among the greatest revolutionaries were Copernicus, who removed the earth from the central location in the universe; Isaac Newton, who provided a mathematical framework for quantifying motion; and Albert Einstein, who fundamentally changed our notion of time and space. The revolutionaries who developed quantum mechanics overthrew our notions of predictability and replaced them with probability and uncertainty.

Revolutionary progress has important implications for physical organic chemistry. The "holy grail" we seek is an understanding of how molecules come together and react. We would like to create a movie where we watch the molecules collide, the bonds break, and new ones form. But quantum mechanics has dashed that dream. Atoms and molecules don't behave like balls on a billiards table. We can't film them colliding and bouncing around the table and falling into pockets, and then rewind that film and play it over and over again.

Furthermore, we can't ever *prove* a reaction mechanism. At best, a mechanism is consistent with all of the experimental data at hand. No single experiment, no set of experiments, can give us that ultimate assurance that this mechanism is correct forever and for all circumstances. However, a single experiment can call a mechanism into question and upset years of work refining a theory. It's the excitement of that next experiment that might shake up the community and bring about that next revolution that drives scientists to continue to dream and develop new experiments, new methods, and new theories.

In the remainder of this book, I will try to convey some of the excitement that surrounds the establishment of chemical theories, the experiments behind those theories, and the tests applied to probe their boundaries.

# 9

# Nucleophilic Substitution Reactions. IV. Further Details

I return to the nucleophilic substitution reaction for a few more details. Let's first address our ability to select for the $S_N1$ or $S_N2$ mechanism, which is done primarily through the choice of solvent.

The solvent is a liquid that dissolves all of the reagents in a reaction. Think of shaking salt into a pot of water—the solid salt crystals disappear by dissolving into the water. Once in solution, the reagents readily move around and randomly bump into each other. We can readily control the rate of the reaction by adjusting the temperature, varying how much kinetic energy the molecules have when they crash into each other. One essential characteristic of a good solvent is that it doesn't react with any of the reactants or products.

The factor that determines *solubility*, the amount of a compound that will dissolve in the solvent (called the *solute*), is their electric polarities. Recall playing with magnets as a child. Magnets have a north pole and a south pole. Opposite poles attract each other, so if you bring the north pole of one magnet near the south pole of a second magnet, the two magnets will snap together. On the contrary, similar poles will repel each other, so when the north poles of two magnets are brought near each other, they repel, they push apart.

This works in an analogous fashion for objects that have a negatively charged end and a positively charged end. The positive end of one object will attract the negative end of another. Many molecules have this property, where one end is positive and the other end has a negative charge. We call this type of molecule *polar*. Molecules that do not have such a charge separation, that do not have a negative end and a positive end are said to be *nonpolar*.

The guiding rule for solubility is "like dissolves like." Polar compounds will dissolve in polar solvents, and nonpolar compounds will dissolve in nonpolar solvents. Since the nucleophilic substitution reactions typically involve polar molecules, if not outright ions, nucleophilic substitution reactions will in general be performed with polar solvents.

In polar solvents, the dissolving process occurs because the dipoles that charge separation in the molecules can align and favorably interact, as

*Thinking Like a Physical Organic Chemist*. Steven M. Bachrach, Oxford University Press. © Oxford University Press 2023.
DOI: 10.1093/oso/9780197640371.003.0009

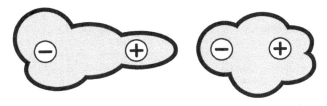

**Figure 9.1.** Favorable dipole–dipole interaction.

depicted in Figure 9.1. The positive end of the electric dipole of the solute molecule can get close to the negative end of the dipole of the solvent, and vice versa.

The strength of this electrical attraction is based on the size of each dipole and the proximity of the dipoles. The greater the magnitude of the dipole, the stronger the attraction to a neighboring dipole. And the closer that the two dipoles can approach each other, the stronger the attraction.

Let's focus on that second factor, getting the dipoles close to each other. The smallest atom is hydrogen, and under appropriate circumstances, the positive end of a dipole will often be located at a hydrogen. This circumstance especially occurs when a hydrogen is bonded to an oxygen or a nitrogen or a fluorine atom. This small hydrogen can then tuck up very close to an oxygen or nitrogen atom on a neighboring different molecule, creating what is referred to as a *hydrogen bond*. This is not a chemical bond. It is much weaker than that, but among the types of interactions *between molecules* (not like a chemical bond, which is *between atoms* within a molecule), hydrogen bonding is the strongest. The hydrogen bond between a water molecule and a methanol molecule is depicted as a dashed line in Figure 9.2a.

Hydrogen bonding is responsible for many of the important properties of water, including its high boiling point and the fact that its solid form (ice) is less dense than its liquid form. Hydrogen bonds are responsible for holding the two strands of a DNA molecule together as a double helix.

Hydrogen bonds can also occur between the positive end of a molecule and a negatively charged atom (an *anion*). This is displayed in Figure 9.2b where each of two methanol molecules forms a hydrogen bond to a chloride ion. Note that a hydrogen bond can't be formed to a positively charged ion (a *cation*); remember that the hydrogen atom of the hydrogen bond carries a positive charge and so it won't get close to a positively charged ion.

All right, so how does solvent affect a nucleophilic substitution reaction? We will examine each mechanism, $S_N1$ and $S_N2$, independently. Let's start with the $S_N2$ mechanism, and draw the reaction coordinate diagram for a

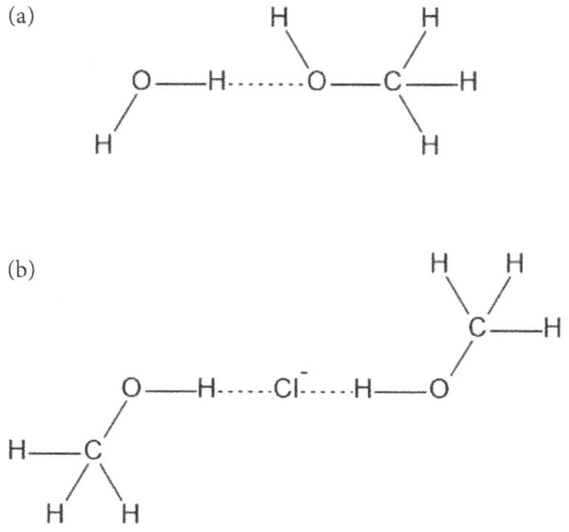

(a)

(b)

**Figure 9.2.** The hydrogen bond (dashed line) in the (a) water–methanol complex and (b) chloride–methanol complex.

generic example taking place in a polar solvent (Figure 9.3). The curve has the expected shape, drawn as the solid line. The reactant lies to the left and the substituted product to the right. As the nucleophile approaches and starts to form a bond to carbon, the leaving group starts to exit, and the overall energy rises. The energy continues to increase as the reaction progresses until the transition state is reached, after which, as the nucleophile continues to get closer to the carbon atom, the energy falls.

Now let's consider the reaction taking place within a hydrogen bonding polar solvent. The energy of any polar or charged species will typically be lower in the hydrogen-bonded solvent than in the polar aprotic (nonhydrogen bonding) solvent. But any concentration of negative charge, like in an anion, will be especially stabilized by the hydrogen bonds formed to the solvent. This means that the reactants, with its $Y^-$ nucleophile, and the products, with its $X^-$ leaving group, will be significantly lowered in energy. On the other hand, the transition state will only be lowered a little because the negative charge is spread out over the forming C–Y bond and the breaking C–X bond. So, in the hydrogen bonding solvent, we've lowered the transition state a little, but we've lowered the reactant energy by a good amount (see the dashed curve and arrow of Figure 9.3, which leads to a larger activation barrier in the (hydrogen bonding) protic solvent than in the polar aprotic (nonhydrogen bonding) solvent. Therefore, if you want to facilitate an $S_N2$ reaction, you should select a polar aprotic solvent.

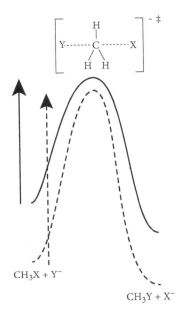

**Figure 9.3.** Reactions coordinate diagram for an $S_N2$ reaction. The solid line depicts the reaction energy profile in a polar aprotic solvent, and the dashed line depicts the reaction energy profile in a polar protic (hydrogen bonding) solvent. The arrows depict their respective activation barriers.

Let's now examine the $S_N1$ reactions in the same fashion. We need only concern ourselves with the first step, the rate-determining step. In this first step, the only chemical change is breaking the C–X bond, leading to the carbocation intermediate and free $X^-$. The reaction coordinate diagram for the reaction in the polar aprotic solvent is shown as the solid curve in Figure 9.4. Switching next to the polar protic (hydrogen bonding) solvent, we look to identify concentrated negative charge. Remember that the nucleophile $Y^-$ is not involved in this first step, so we need not consider it. The reactant will be stabilized by a small amount in the polar protic solvent, but significant stabilization will be had in the product, having created the free $X^-$ species that makes strong hydrogen bonds to the solvent. The transition state involves stretching the C–X bond, putting more negative charge on the leaving group. The buildup of negative charge on the leaving group will be stabilized by hydrogen bonds—not as much as with the free $X^-$ in the product, but more so than for the reactant. This means that the activation barrier is lower in the protic solvent than in the aprotic solvent (see the dashed curve in Figure 9.4). So, an $S_N1$ reaction will be

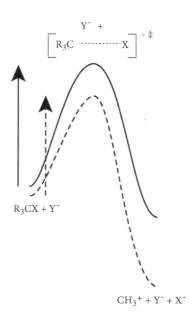

**Figure 9.4.** Reactions coordinate diagram for the first step of an $S_N1$ reaction. The solid line depicts the reaction energy profile in a polar aprotic solvent, and the dashed line depicts the reaction energy profile in a polar protic (hydrogen-bonding) solvent. The arrows depict their respective activation barriers.

facilitated by running it in a polar protic solvent. This is different from the $S_N2$ case, which is facilitated by a polar aprotic solvent.

Organic chemists really like such situations. They want to be able to start with a specific set of reagents and make specific products. They also want to be able to make small adjustments and completely control changes in the outcomes. So, if we need to perform a nucleophilic substitution on a primary system, we'll choose conditions for an $S_N2$ reaction (like selecting a polar aprotic solvent), but if we need to perform a nucleophilic substitution on a tertiary system, we'll use $S_N1$ conditions (polar protic solvents). This means, for example, that if we have a molecule with both primary and tertiary reactive centers, we can selectively perform the reaction at just one of those places and keep the other one untouched, as represented in Figure 9.5.

The next topic is to revisit the stereochemistry of the $S_N2$ reaction. The example reaction presented earlier (see Figure 7.11) shows an $S_N2$ reaction taking place at a secondary carbon. It is reasonable to ask about the stereochemistry of this reaction at a primary carbon, which is in fact the type of system most prone to an $S_N2$ mechanism.

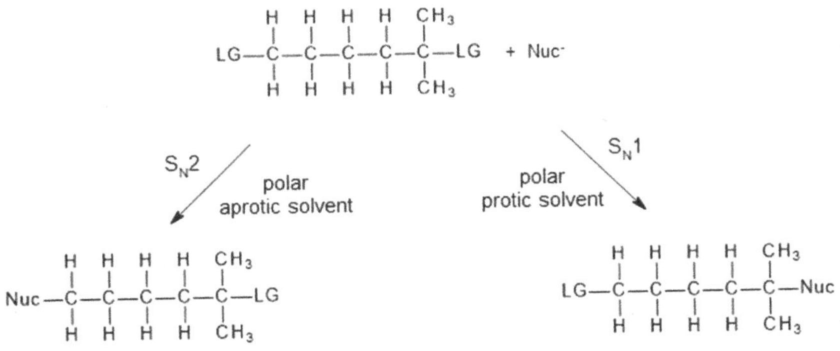

Figure 9.5. Selective substitution reactions.

The key element for detecting stereoinversion in an $S_N2$ reaction is to have a chiral center that can invert during the course of a reaction. A simple way to identify a chiral carbon is one that has four different groups attached to it. One of those groups will be the leaving group, and one will be a carbon chain—and the other two will be hydrogen atoms. That means the carbon will be achiral, since only three (not four!) different groups are attached to carbon.

So, how can we make a primary carbon chiral? To answer this question, we have to talk about what's in an atom. The nucleus has two different components: protons and neutrons. Protons carry a positive charge and neutrons carry no electric charge. An element is defined by how many protons are in the nucleus. All carbon atoms will have six protons, and all hydrogen atoms will have a single proton. For most elements, the majority of their atoms will also have the same number of neutrons. Most carbon atoms will have six neutrons. We will call this a C-12 atom since it has six protons and six neutrons. Some carbon atoms, however, will have seven neutrons, called C-13, and even more rarely we find a carbon atom with eight neutrons, C-14. Atoms of the same element, meaning that they all have the same number of protons but a different number of neutrons, are called *isotopes*.

Perhaps the most famous isotopes are those of uranium: U-235 and U-238. Both have 92 protons; that's what makes them uranium. The U-238 isotope has 146 neutrons and is radioactive. The other isotope, U-235, has 143 isotopes, and not only is it radioactive, it is the fuel for an atom bomb. U-238, the more common isotope, is not a fuel for an atomic reaction; in fact, it retards the atomic reactions. In order to make an atomic bomb, or a fission nuclear reactor, you have to separate the U-238 from the U-235. The enriched U-235 can then be used as the fuel in a nuclear reactor, and U-238, often called *depleted uranium*, can be used for other non-nuclear purposes. Fortunately, separating these two isotopes is a difficult process!

Most hydrogen has a nucleus with just a single proton. A small amount of hydrogen, however, has a nucleus with one proton (that's again what makes it hydrogen) and one neutron, H-2. Chemists have given this isotope its own name, deuterium with symbol D, but keep in mind that chemically it behaves just like a regular hydrogen atom.

That single small difference of one neutron in deuterium and no neutron in a normal hydrogen make them different enough to be useful. What happens if we replace one of the hydrogens on a primary carbon with a deuterium, for example as in the butanol molecule shown in Figure 9.6?

(Let's take a very short interlude to discuss some simplifications in drawing chemical structures, simplifications that all chemists use in their everyday practice. Carbon is the ubiquitous element in organic chemistry, so let's stop writing out the letter C; instead, we will just draw the bonds as lines, and the end of a line with no atomic symbol will be understood as a carbon atom. Hydrogens are also ubiquitous and are often of minimal chemical consequence. Let's just omit drawing them in when they are attached to carbon. Since we know that all carbon atoms will make four bonds, if we count the number of bonds to a particular carbon atom, the number of missing (undrawn) hydrogens will be enough to make the carbon have a total of four bonds. So the end of the line at the left of the first molecule is a carbon atom, and since it has only one line drawn to it, there are also three hydrogens bonded to that terminal carbon. The next carbon over to the right is the first vertex, and since there are two C–C bonds, it must be bonded to two hydrogen atoms.

The two deuterium-substituted butanol molecules in Figure 9.6 are identical in all aspects, but they are not superimposable. They are mirror images of each other; they are chiral. Each of these molecules will be optically active, and that difference of how they interact with polarized light is their sole physical difference. As enantiomers, their chemical behavior will be identical.

Andrew Streitwieser, an important physical organic chemist during the heyday of the field, performed a study of the nucleophilic substitution reaction on a slightly modified deuterium-labeled butanol. This molecule differs

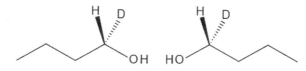

**Figure 9.6.** The enantiomers of deuterated 1-butanol.

from butanol by having a better leaving group, thus making the reaction proceed faster. Examination of the product shows essentially complete inversion of the stereochemistry, as is expected for an $S_N2$ reaction. This experiment definitively implicated (since the $S_N1$ reaction is never taking place at a primary carbon) the notion of a backside attack through a single chemical step.

We have drawn the backside attack of the nucleophile within the $S_N2$ mechanism with a linear arrangement of the nucleophile, carbon atom, and leaving group. Certainly, having them arranged in this manner keeps the nucleophile and leaving group from running into each other. Are there limitations on this attack angle? Can the nucleophile approach off this axis and a reaction still occur?

At first blush, this seems to be an impossible task. How can we control the path that the nucleophile takes as it approaches the carbon? How can we know what that path is, especially in light of the limitations imposed by quantum mechanics?

The solution, developed by the Swiss organic chemist Albert Eschenmoser, most noted for his work in the synthesis of vitamin $B_{12}$, is incredibly elegant. I consider this experiment to be one of the most clever in physical organic chemistry for its simplicity of design, and its powerful, clear interpretation. I think it is one of the more overlooked gems in our field.

The experiment goes by the name *endocyclic restriction test*. The term *endocyclic* means inside a ring. Eschenmoser recognized the prevalence of six-membered rings throughout organic chemistry and the many examples of transition states that take place through six-membered rings. (We will see many examples of this in Chapter 17, which discusses pericyclic reactions.) He imagined setting up a nucleophilic substitution reaction that might be constrained (restricted) to take place within a six-membered ring. He could accomplish this by tying the nucleophile, the carbon under attack, and the leaving group all within one larger molecule. That geometric constraint would hold the angle of attack somewhere near 120°, and decidedly not linear.

How did Eschenmoser create such a constrained structure? The answer is the molecule shown in Figure 9.7. Looking at the reactant, we note that the nucleophile is the carbon with a negative charge, the carbon under attack is the bold $CH_3$ group, and the leaving group is the oxygen in bold. This substitution reaction involves making the C–C bond and breaking the C–O bond. The carbon under attack is a methyl group, and it will react only under the $S_N2$ mechanism.

In Figure 9.7, each of the six-membered rings (called a *phenyl* ring) is planar, and the angles about each carbon are very close to 120°. This top phenyl ring, as well as the angles about the sulfur and oxygen atoms, help to create what

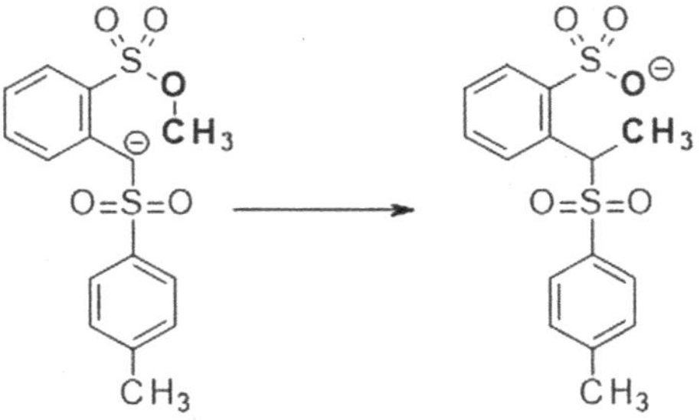

**Figure 9.7.** Substitution reaction in Eschenmoser's endocyclic restriction test.

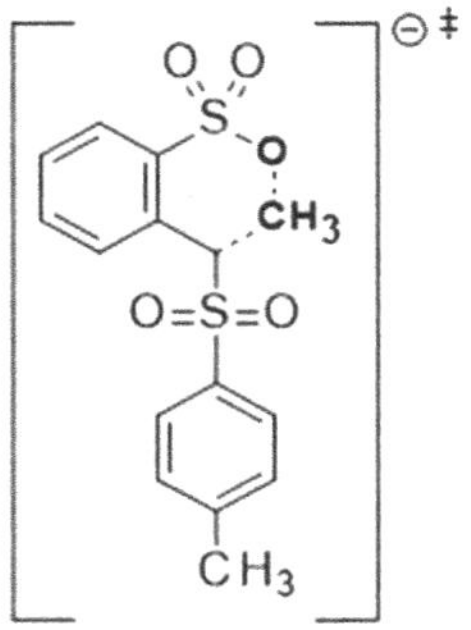

**Figure 9.8.** Intramolecular (unimolecular) transition state for the endocyclic restriction test.

looks to be an ideal possibility of a cyclic transition state. This would be an *intramolecular* reaction involving just this one molecule, traversing through a transition state that looks like the one depicted in Figure 9.8.

But what if a backside attack in the $S_N2$ mechanism really needs to have a nucleophile–carbon–leaving group relationship that is nearly linear? Arguments based on quantum mechanics do suggest that the attack needs to be nearly linear for optimal electron distribution in the transition state. It is possible for the reaction to occur in an *intermolecular* fashion. In this case, shown in Figure 9.9, two molecules come together. The first molecule acts as the nucleophile, as the carbon with a negative charge attacks the carbon in bold face on the second molecule. The leaving group, the bold face oxygen, is also on the second molecule.

Notice that this sets up favorably for the intramolecular (unimolecular) process. Entropy dramatically favors the unimolecular process over the

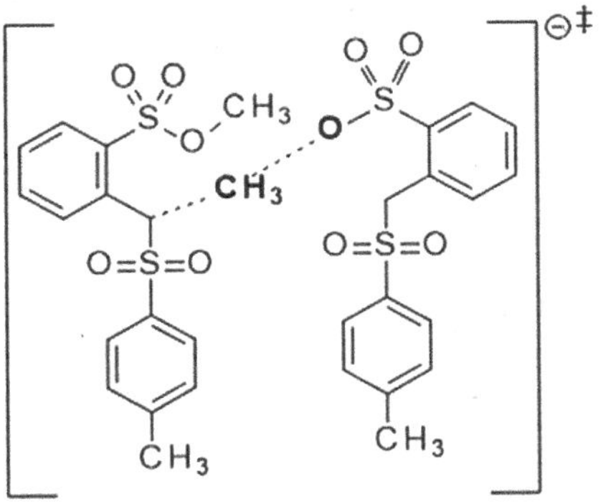

**Figure 9.9.** Intermolecular (bimolecular) transition state for the endocyclic restriction text

bimolecular process. In the unimolecular process, the reactive ends need to find each other, and that doesn't require too much ordering of the geometry since the ends are positioned close by. In the bimolecular process, two independent molecules have to come nearby in proper orientation for the reaction to take place, necessitating a significant loss of entropy.

How can we distinguish which path is operational? What we need to do is somehow "color" parts of the molecules to keep track of where they start and where they end up. Isotopic labeling is an excellent way to do just that: it provides a way to tag a portion of the molecule without affecting its chemistry.

Eschenmoser devised a crossover experiment with clever isotopic labeling. In a *crossover experiment*, two nearly identical species react within the same reaction vessel. This provides for each compound to react independently, where the products derive from the reaction of just one of the compounds, or for the two compounds to react with each other, in which case the product has a component from each compound (the crossover product). Let's see how this was done in Eschenmoser's endocyclic restriction test (Figure 9.10).

Each of the reactant molecules, **A** and **B** are labeled in two places, the two $CH_3$ groups, called the *methyl* groups. The methyl on the bottom tags the nucleophile, while the methyl connected to the oxygen tags the center where the substitution takes place. These two molecules are distinguished by having only hydrogens on the methyl groups (**A**) or only deuterium on the methyl groups (**B**).

Suppose we mix equal amounts of **A** and **B** in a reaction flask. What we will obtain depends on whether the reaction proceeds unimolecularly (through

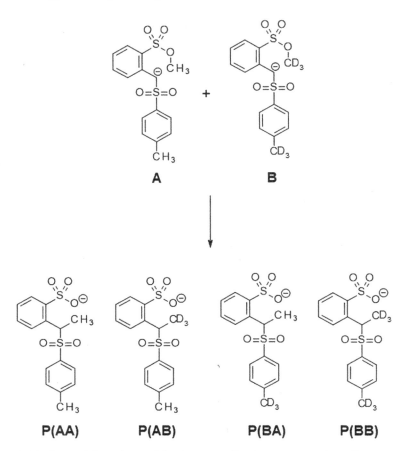

**Figure 9.10.** Potential products of the deuterated Eschenmoser endocyclic restriction test.

the transition state shown in Figure 9.8) or bimolecularly (through the transition state shown in Figure 9.9). In the unimolecular case, there is no crossover, so no scrambling of the labels takes place. The products will be either **P(AA)**, which comes from the reaction of **A**, or **P(BB)**, which comes from the reaction of **B**. We would then expect to see equal amounts of these two products, and none of the other two products.

Suppose instead, that the reaction takes place bimolecularly. Compound **A** can be the nucleophilic portion, and it could react with another **A** molecule or equally likely with a **B** molecule. Similarly, **B** might act as the nucleophile and react with another **B** molecule or equally likely with an **A** molecule. So, all four products would be expected, and in equal amounts.

So, what was the outcome of the endocyclic restriction test? The four products were found, in equal amounts. This means that the reaction went bimolecularly, *even though that pathway is entropically disfavored*. The implication is that the unimolecular path is uncompetitive because of the constrained angle of approach that the nucleophile can make. The $S_N2$ pathway requires true backside attack, with the nucleophile, carbon, and leaving group lined up.

# 10

# Elimination Reactions. I. $E_1$ and $E_2$

It's about week five into a typical college organic class, and you've made it through nucleophilic substitution reactions. You have some confidence that you can distinguish when a reaction will proceed by the $S_N1$ mechanism and when it will proceed by the $S_N2$ mechanism. You wish perhaps that the world was a little cleaner, but you're ready to accept some small level of disorder.

You sit down in my classroom, and I start my lecture by drawing the reaction in Figure 10.1 on the board.

This is an especially bitter pill to swallow in an 8:30 a.m. lecture. It was difficult enough to grasp the idea that nucleophilic substitution reactions can proceed by two different mechanisms. But this indicates that a *competing reaction* is taking place. The first product is the substitution product from an $S_N1$ process since this is a tertiary system. The second product comes about through removal of chlorine from a carbon and a hydrogen on a neighboring carbon, resulting in a double bond between these two carbon atoms. That's not a substitution reaction!

The next reaction (Figure 10.2) I write doesn't make things any better. Again, that first product is fine; it's the expected $S_N2$ product this time. But that second product also comes from loss of a hydrogen and a bromine on neighboring carbon atoms.

Can it possibly be that reagents might react in different ways and lead to multiple different products? Unfortunately, the answer is yes, and this is not an unusual occurrence. This complication of competing reactions certainly makes learning organic chemistry much more challenging.

So what's going in these two examples? The second product in both reactions contains a carbon–carbon double bond, molecules that we call *alkenes*. The reaction involves loss of a hydrogen atom and some leaving group on an adjacent position, creating the double bond between the two carbon atoms that lost the groups. This is named an *elimination reaction*, for obvious reasons.

As with nucleophilic substitution reactions, there are two different reaction mechanisms for elimination reactions, $E_1$ and $E_2$. You can likely surmise that $E_1$ is a unimolecular process and $E_2$ is a bimolecular process. I don't want to go through all of the details that underlie these two mechanisms. In this and the following chapter, I'll just walk through the mechanisms and describe some

*Thinking Like a Physical Organic Chemist*. Steven M. Bachrach, Oxford University Press. © Oxford University Press 2023. DOI: 10.1093/oso/9780197640371.003.0010

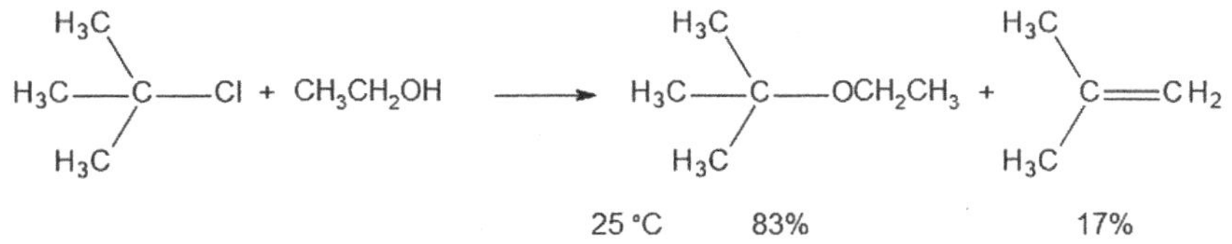


**Figure 10.1.** Reaction of t-butylchloride with ethanol.

$$CH_3CH_2Br + CH_3CH_2O^- \longrightarrow CH_3CH_2OCH_2CH_3 + H_2C{=}CH_2$$

50% 50%

**Figure 10.2.** Reaction of bromoethane with ethoxide.

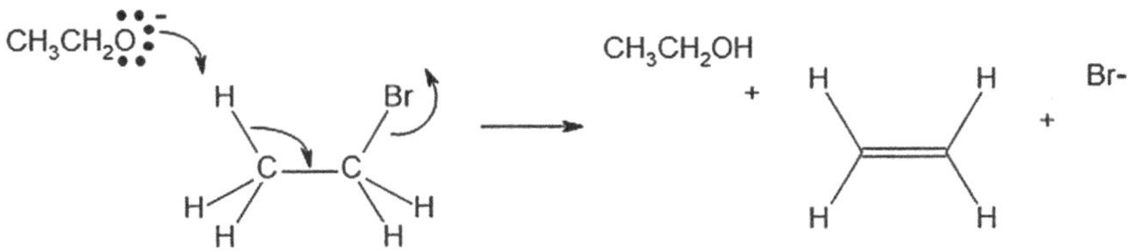


**Figure 10.3.** Mechanism for the $E_2$ reaction of bromoethane and ethoxide.

experiments that elaborate a few interesting aspects of these reactions, along with a major concept of chemical structure in organic chemistry.

The mechanism for the $E_2$ reaction is shown in Figure 10.3. First, I need to explain the curved arrows in this diagram. These arrows indicate the movement of electron pairs, showing the making and breaking bonds. The arrows originate with some pair of electrons. These can be the electrons in a bond, one of the pairs of electrons in a double or triple bond, or a pair of electrons that are unshared, residing on just a single atom. The head of the arrow points to where the electrons are going. Where might a pair of electrons head toward? The obvious (and correct) answer is toward positive charge! An atom that carries some positive charge can be the electron sink, the place where electrons will wish to move. But the electron sink might also be at an atom that wants to gain electrons. Elements that are eager to gain electrons to fill their octet lay over on the right side of the periodic table, such as fluorine, chlorine, bromine, iodine, and oxygen. (You might remember from some earlier chemistry class that these atoms are characterized as *electronegative.*) Bond breaking is indicated by an arrow that starts at a bond and heads away, while bond making is indicated by an arrow that points between two atoms. A reaction mechanism can then be drawn with one or more arrows to represent the bond changes through electron pair movement. We call this *arrow pushing*, a concept first articulated by the British chemist and Nobel Laureate Sir Robert Robinson. It is not to be understood as "this is *the* motion of electrons in a reaction." Remember, quantum mechanics *prohibits us from knowing that*! Rather, this arrow pushing is a representation, a model, that allows us to make predictions of how a reaction might take place, and then design experiments to test that prediction.

What do the arrows for the E$_2$ mechanism in Figure 10.3 tell us? The electron source, the start of the arrows, is at the oxygen of ethoxide (CH$_3$CH$_2$O$^-$). This oxygen has three lone pairs, three unshared pairs of electrons. One of these pairs will move to attack a hydrogen of the terminal carbon of bromoethane (CH$_3$CH$_2$Br), leading to the creation of an O–H bond. This first arrow "pushes" the second arrow, which indicates that the C–H pair of electrons moves to form the second bond of the incipient double bond. That movement "pushes" the third arrow, indicating that the pair of electrons used in the C–Br bond moves to the bromine atom, cleaving that bond. Note that the electrons end up in the sink, with the very electronegative bromine atom. All of these changes take place within one chemical step, through a single transition state, analogous to what happens in the S$_N$2 reaction.

The mechanism for the E$_1$ reaction is shown in Figure 10.4. In the first step, the C–Cl electron pair moves to the chlorine, breaking the bond and creating a tertiary carbocation. In the second step, one of the lone pairs on the oxygen of ethanol (CH$_3$CH$_2$OH) moves to create a bond with a hydrogen, pushing the C–H bonding electron pair to move to the carbocation (the electron sink), making the double bond. This is a two-step reaction, just like in the S$_N$1 reaction.

Let's take a look at the reaction shown in Figure 10.2 and consider the S$_N$2 and E$_2$ mechanisms that are in competition. For the S$_N$2 pathway, ethoxide (CH$_3$CH$_2$O$^-$) acts as the nucleophile, attacking the partially positive carbon

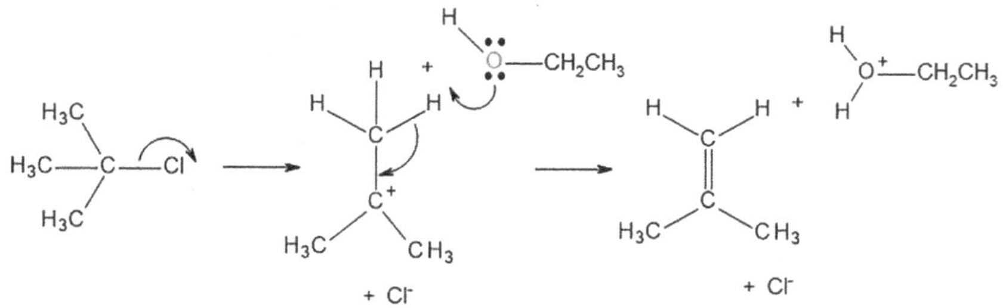

Figure 10.4. Mechanism for the E$_1$ reaction of t-butylchloride with ethanol.

that is bonded to bromine. That same molecule, ethoxide, can just as well attack the partially positive hydrogen atom to start the $E_2$ reaction. In this latter case, ethoxide is acting as a *base*.

Acid–base reactions are fundamental to chemistry. They are responsible for acidification of the ocean, discoloration of buildings and statues through acid rain, unclogging drains using lye, manufacturing soap, the tangy taste of vinegar and lemon juice, and many other uses. Acids and bases can be defined in a few ways. The oldest and most common definition of these reactions comes from the Swedish chemist Svante August Arrhenius, for which he was awarded the Nobel Prize in 1903. An Arrhenius acid is a compound that produces $H^+$ when dissolved in water, and an Arrhenius base is a compound that produces $OH^-$ when dissolved in water. An extension of this notion removes the need for water as the solvent. Developed independently by the Danish chemist Johannes Brønsted and the English chemist Thomas Lowry, a Brønsted-Lowry acid is a compound that gives up, or donates, an $H^+$ and a Brønsted-Lowry base accepts, or bonds to, an $H^+$.

In elimination reactions, the base picks up the $H^+$ from the substrate. In substitution reactions, that same molecule acts as the nucleophile, attacking the partially positive carbon atoms. It comes down to the difference in attacking positive charge on a hydrogen versus positive charge on a carbon. Molecules can often display this Janus-like behavior, able to react in two different ways. Which process dominates depends on the rather subtle interplay of effects that we need not concern ourselves with here—though it is a necessary component of an organic chemistry class. Be grateful that I am not quizzing you on this.

Instead, I will next explore the regiochemistry of elimination reactions. I will introduce that concept with the example reactions shown in Figure 10.5.

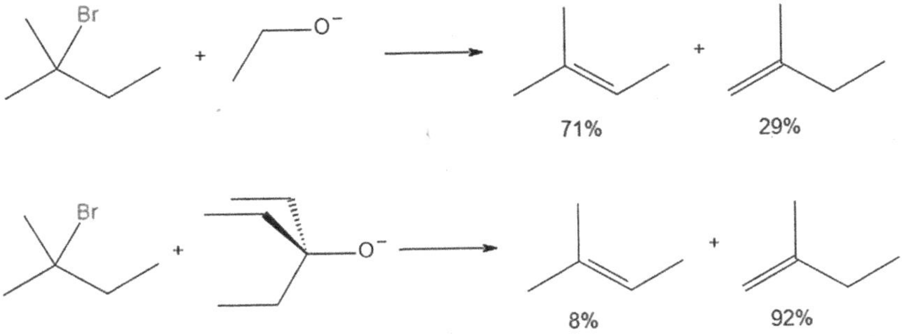

Figure 10.5. Reaction of 2-bromo-2-methylbutane with a small, base or with a bulky base.

In these two examples, which differ in the base used, two different alkenes are made as the elimination products. Substitution likely also occurs, but for now we are focusing on just the elimination products.

The first alkene results from loss of the hydrogen on the third carbon, while the other product comes by loss of a hydrogen from the first carbon. That's why the term *regiochemistry* applies; we are distinguishing which region of the molecule will react. The most useful distinction of these two alkenes is their degree of substitution. The first alkene has three substituents, three carbon-containing groups, attached to the two carbon atoms of the double bond. The second alkene has two substituents attached to the two carbon atoms of the double bond.

Extensive studies of the stabilities of alkenes identified a simple trend: the more substituted the alkene, the more stable it is. So, of the two alkenes made in the reactions shown in Figure 10.5, the first alkene is more stable than the second one. If we consider that the universe tends to move to the lowest energy structure available—think of the rock rolling down a hill—then we would expect that the first, more stable alkene would be the major product. We refer to the most stable product as the *thermodynamic product*. The top reaction in Figure 10.5 predominantly yields the thermodynamic product.

However, the second reaction flips the preference; the less stable alkene is the major product. What has changed here from the first reaction? In order to make the thermodynamic product the first alkene, the base has to remove a hydrogen from the third carbon. That's easily done by a small base, like the one used in the top reaction, but the base employed in the bottom reaction is quite bulky. It will not readily squeeze into the interior to get to the hydrogen on the third carbon. Rather, it will be much easier for this big base to remove the hydrogen from the end of the molecule, leading to the second alkene.

In this case, what's easiest to accomplish determines the product. That implies a rate, that is, how quickly something can be done. A reaction with a lower barrier will run more quickly and will produce more product than a competing reaction with a higher barrier. Plucking off the terminal hydrogen requires less loss of entropy, less arranging of the base and the substrate, than will be the case from attacking the interior hydrogen. The base and substrate will likely push into each other in ways that will increase their energy when going after the interior hydrogen, which won't happen when pulling off the terminal hydrogen. So, the second, less stable product is formed faster, and we refer to it as the *kinetic product*. By playing with the base, organic chemists can alter the preference for the thermodynamic and kinetic products. We will see other examples of this trade-off, this competition, between kinetic and thermodynamic products in subsequent chapters, where I will discuss this competition in more detail.

For the next topic, I will need to briefly discuss molecular orbitals. Orbitals are regions of space where we are likely to find electrons. Remember that quantum mechanics only allows us to talk about probabilities, so we can't point to a location and say, "here is where the electron can be found." We can't treat an electron like it's a classical particle, like a billiard ball. Orbitals are as good as it gets, and we can only say "in this region of space, this *orbital*, we will find the electron 95% of the time."

I want to take a look at the second bond of a typical C–C double bond. We call this a *π-bond*. It is constructed of the combination of a p-orbital on each of the neighboring carbon atoms. The p-orbital looks like a dumbbell. The top and bottom lobes of the dumbbell have different mathematical signs (one is positive and one is negative). The π-bond is made from the side-by-side arrangement of these neighboring p-orbitals, as shown in Figure 10.6.

That side-by-side alignment of the p-orbitals suggests that the breaking C–H and C-leaving group bonds are aligned in the same plane, or *periplanar*, in order to form the π bond. These two bonds might be on the same side (*syn-periplanar*) or on opposite sides (*anti-periplanar*), as shown in Figure 10.7.

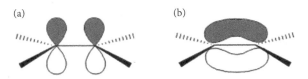

**Figure 10.6.** Orbital representation of the π-bond: (a) constituent p-orbitals and (b) resulting molecular orbital.

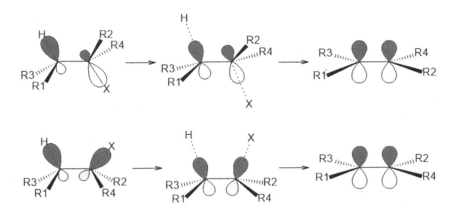

**Figure 10.7.** Orbital progression in the anti-periplanar (top) and syn-periplanar (bottom) pathways of the E$_2$ mechanism.

The leaving group is represented here as X, which is often the abbreviation we use for a halogen, a frequently employed leaving group. Elimination either by the syn-periplanar or anti-periplanar route insures optimal arrangement of the orbitals throughout the reaction coordinate.

So, how can we test if these pathways are in fact being followed, recognizing, as always, that quantum mechanics forbids us from knowing *exactly* what takes place? Let's take a closer look at the two possible pathways shown in Figure 10.7. First, recognize that the two reactants are the exact same molecule, shown to differ only in how the groups are arranged in space. That difference is simply arrived at by holding the left carbon fixed, and rotating the right carbon 180° about the C–C bond. Molecular arrangements that differ just by the rotation about one or more bonds are called *conformational isomers*, or *conformers* for short.

The two conformations of the reactant in Figure 10.7 are displayed in a *sawhorse representation*. Viewing the bottom conformation, it has two sets of legs, one on the left and one on the right. In each set, one leg projects in front of the page and one leg is behind the page. These legs, along with a top defined by the central C–C bond, appear like a sawhorse employed in construction or woodworking projects.

Conformers are typically close in energy, but they can often be distinguished. To aid in seeing how this is done, we make use of the *Newman projection*, a visualization tool developed by the American chemist Melvin Newman. Imagine positioning your eye to look straight down the C–C bond of a molecule, such as in Figure 10.8. Looking at the molecule this way, we

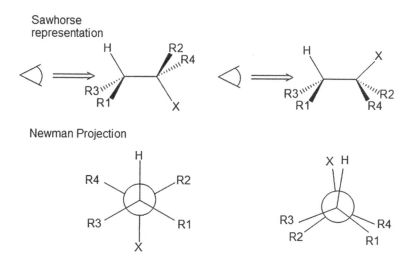

**Figure 10.8.** Sawhorse representation (top) and Newman Projection of the staggered (left) and eclipsed (right) conformations.

note that the left carbon is now in front, with the right carbon oriented behind it. We will use a dot to represent the carbon in front and a circle to represent the carbon in back. Bonded to the front carbon are the hydrogen atom and the R1 and R3 groups. The hydrogen atom is straight up, the R1 group is down to the right, and the R3 group is down to the left. Draw these connected to the central dot, representing the front carbon atom. Bonded to the back carbon are the leaving group X and the R2 and R4 groups. The leaving group is straight down, the R2 group is up to the right, and the R4 group is up to the left. Draw these groups connected to the circle, representing the back carbon. Note that the six groups bonded to the two carbons lie between each other. We refer to this conformation as *staggered*.

Now let's examine the other conformer, where the right carbon has been rotated by 180°. This is the molecule on the right in Figure 10.8. Again, we will look down the C–C bond, directly at the left carbon, with the right carbon behind it. Since the front carbon is unchanged in this conformer from the first one we looked at, the Newman representation will again have the hydrogen straight up, the R1 group down to the right, and the R3 group down to left. Bonded to the back carbon is the X group, now straight up, the R4 group down to the right, and the R2 group down to the left. The X group should be drawn right behind the H atom, but for clarity, we move it a bit over, and do the same for the R2 and R4 groups. The three groups attached to the front carbon lie right in front of the three groups attached to the back carbon; we refer to this as the *eclipsed* conformation.

So, what's the energy difference between the staggered and the eclipsed conformations? In the eclipsed conformations, the groups attached to the two carbon atoms get very close and can bump into each other. In the staggered conformation, the groups are farther apart and are less likely to get into each other's space. Thus, the staggered conformation is a bit more favorable (lower in energy) than the eclipsed form.

Since the staggered form is more stable than the eclipsed, one might then expect that the anti-periplanar pathway is preferred to the syn-periplanar one. The anti-periplanar pathway also keeps the two partially negatively charged groups—the incoming base and the exiting leaving group—far from each other.

OK, we expect the anti-periplanar reaction pathway, but how can we tell if that route actually occurs? How can we distinguish it from the syn-periplanar pathway? Let's go back and examine the products of the two reactions in Figure 10.7. Remember that alkenes are planar because of the π-bond. That π-bond inhibits rotation about the C–C double bonds; rotation would force the π-bond to break, as the two p-orbitals would become unaligned. In the

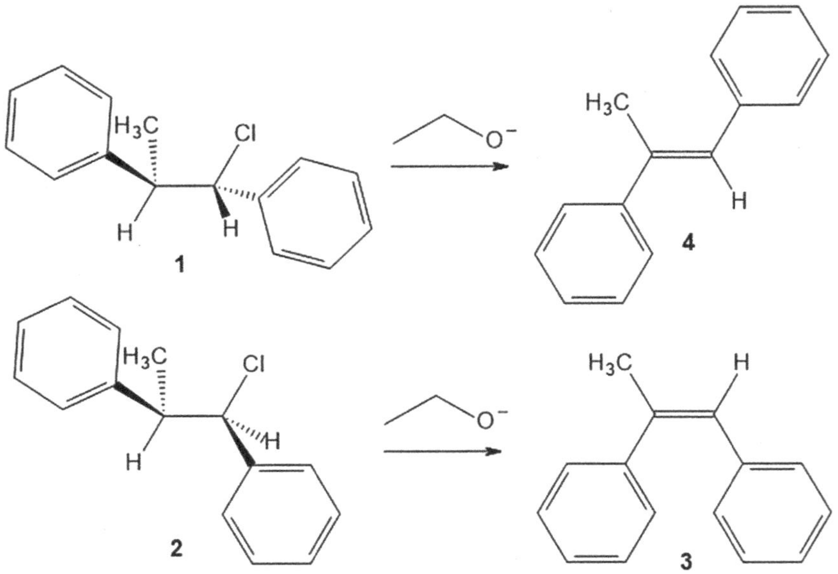

**Figure 10.9.** Cram's elimination reactions identifying the anti-periplanar pathway.

product of the anti-periplanar pathway, the R1 and R2 groups are on the same side of the double bonds, or what we call a *cis* arrangement. But the product of the syn-periplanar path has the R1 and R2 groups on opposite sides of the double bond, called a *trans* arrangement.

Donald Cram, an American chemist who shared the 1987 Nobel Prize for unrelated work, cleverly designed two related molecules that were then exposed to base (Figure 10.9). Both reactants, 1 and 2, could in principle undergo an elimination reaction to give the cis and trans alkenes 3 and 4. I have drawn the staggered conformations of 1 and 2 set up for anti-periplanar elimination.

The reaction of 1 with base gives just the trans alkene 4. That might suggest that only the anti-periplanar pathway is executed. However, the trans isomer is more stable than the cis isomer: those two large ring substituents bump into each other in the cis isomer 3 but are far apart in the trans isomer 4. Perhaps the elimination reaction of 1 simply yields the most stable product, the thermodynamic product? This is where the experimental design is so clever. The anti-periplanar elimination product for the reaction of 2 is the cis isomer, the less stable possible product. Yet, the reaction of 2 produces only 3, the less stable isomer. The fact that each reaction produces a single alkene isomer and that each of these is the one expected from the anti-periplanar pathway is strong supporting evidence that the $E_2$ mechanism proceeds by anti-periplanar loss of the neighboring groups.

Another finely crafted experiment adds weight to the anti-periplanar pathway, but telling this story requires a diversion in the notion of conformational analysis of six-membered rings. I address this topic in the next chapter.

# 11

# Elimination Reactions. II. Conformation and Stereochemistry

We have already seen examples of six-membered rings in this book, and they are ubiquitous throughout organic chemistry. They are found in sugars, in steroids, in pharmaceuticals, and in plastics. They are also present in petroleum and in products derived from petroleum.

There are two common six-membered rings in organic chemistry. The first is cyclohexane, which contains six carbons, all of which are connected by single bonds (see Figure 11.1). The second is benzene, which also contains six carbon atoms, but they are connected by three single bonds and three double bonds in alternating positions about the ring. The cyclohexane-like structure is what occurs in diamond. The benzene-like structure is found in graphite.

Why are six-membered rings so prevalent? To answer that question, we have to delve into their structures. I'll discuss benzene later in Chapter 16 and the cyclohexane structure here. To do that, I need to introduce the notion of *conformational analysis*, the study of the three-dimensional shapes of molecules that differ by rotations about C–C single bonds. Our entry into conformational analysis was with the discussion of the staggered and eclipsed isomers in the previous chapter. Let's extend that discussion now to cyclohexane.

At first blush, we might consider cyclohexane to have the structure of a hexagon, a perfectly symmetrical, planar arrangement, as shown in Figure 11.2. This was the structure proposed by the German chemist Adolf von Baeyer, one of the founders of modern organic chemistry and the recipient of the 1905 Nobel Prize in Chemistry. Baeyer argued for this structure because of the absence of isomers seen with cyclohexane rings with a single substituent. However, based on just what I have discussed in this book so far, we can see that this planar hexagonal structure is very problematic.

Ideally, each of the carbon atoms in cyclohexane will have a near tetrahedral arrangement of its bonds. The angle about a perfect tetrahedral center is 109.47°. The C–C–C angle in a planar hexagonal molecule should be 120°. That's quite a bit of angular distortion, requiring some creation of strain in the molecule. And keep in mind that would happen at each of the six carbon atoms!

*Thinking Like a Physical Organic Chemist.* Steven M. Bachrach, Oxford University Press. © Oxford University Press 2023.
DOI: 10.1093/oso/9780197640371.003.0011

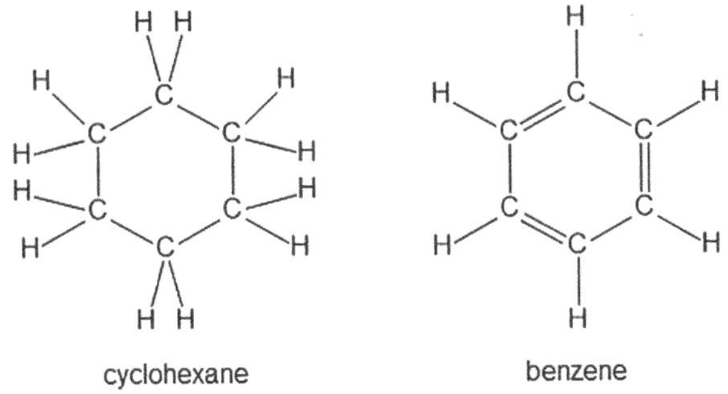

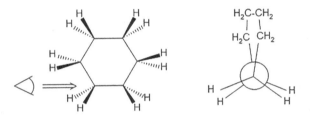

**Figure 11.1.** Structure of cyclohexane and benzene.

**Figure 11.2.** Newman Projection of a planar cyclohexane structure.

Next, let's examine the conformation about each of the C–C bonds. Look down a C–C bond and draw the Newman Projection of this planar hexagonal structure, as depicted in Figure 11.2. Connected to the front carbon are two hydrogens, one down to the right and one down to the left. Straight up is the C–C bond to the $CH_2$ group that continues on to form the ring. Connected to the back carbon are two hydrogens, one down to the right and one down to the left. The third group connected to the back carbon is the $CH_2$ and it is straight up. In other words, this is an eclipsed arrangement that we know is less energetically favorable than the staggered isomer. Again, this less favored conformation is happening at all six bonds. Clearly, if the molecule could distort in some way to remove the eclipsing interactions and decrease each of the C–C–C angles, cyclohexane would readily adopt that new, significantly more stable conformation.

The solution was developed by Herman Sachse in 1890. He devised a non-planar structure whose C–C–C angles were all near the ideal tetrahedral arrangement with staggered arrangements about each C–C bond. He developed this model solely using trigonometry, but his argument fell on deaf ears, largely because of the dense trigonometric treatment made within his article and an absence of the language familiar to chemists.

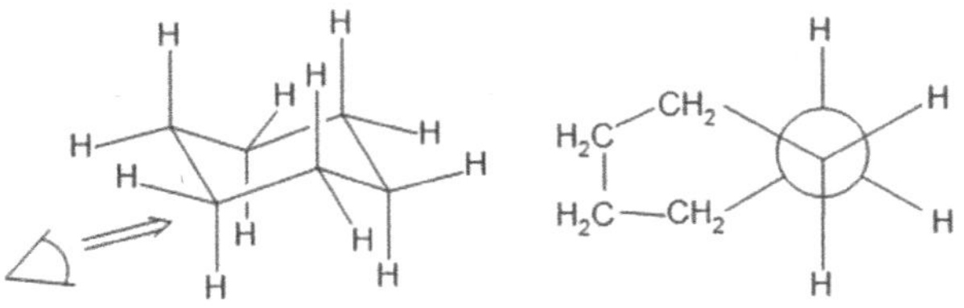

**Figure 11.3.** Newman Projection of the chair conformation of cyclohexane.

You are probably familiar with scientific explanations that seem to be cobbled from words foreign to any human being. This is a longstanding communications problem that scientists have had to contend with when connecting to the lay audience—and it is part of what I am trying to rectify with this book! This problem exists even within scientific communities: arguments are made in terse prose that even experts can often find difficult to decipher. A famous example is the development of quantum electrodynamics in the 1940s and 1950s. Julian Schwinger developed the mathematics of renormalization, but almost no one adopted it, largely because of his use of unfamiliar mathematics couched in opaque explanations. At about the same time, Richard Feynman developed a diagrammatic approach to doing these computations, diagrams that are easily understood and readily translated into mathematics using simple rules. The Feynman Diagram became the *lingua franca* of quantum computations. Later on, Freeman Dyson wrote a famous article that connected Schwinger's and Feynman's approaches, written in a way that made the computations understandable to physicists. Scientific communication remains a skill that eludes many practitioners, leading to unfortunate rediscovery of work.

So what was Sachse's take on the structure of cyclohexane? By allowing the carbon atoms to move out of a single plane, the molecule can adopt the conformation shown in Figure 11.3. This conformation can be thought of as resulting from distorting the planar structure by pulling one carbon downward and the carbon on the opposite side upward. The conformation is given the name *chair*. If you let your imagination run, it looks like a recliner, with your head leaning on the back at the left side and your legs relaxing on the downward incline portion on the right (see Figure 11.4).

A Newman Projection down a C–C bond in the chair conformation of cyclohexane is presented in Figure 11.3. The two hydrogen atoms connected to the front carbon are positioned straight down and up to the right. The $CH_2$ group is up to the left. For the back carbon, the two hydrogens are straight up and down to the right, with the $CH_2$ group down to the left. This is a staggered conformation. That solves the eclipsed bond problem.

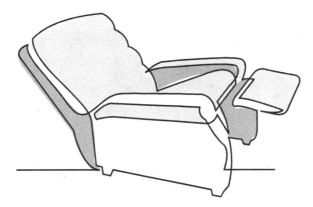

**Figure 11.4.** A recliner as the inspiration of the chair conformation.

(a)

(b)

**Figure 11.5.** Molecular model of cyclohexane. Note the interchange of the axial and equatorial hydrogens—the white and gray sticks—upon the ring flip.

The C–C–C angles in the chair conformation are all around 109.5°, removing the angle strain. This can be seen in the models show in Figure 11.5. These models are constructed with the bond angles fixed at the ideal tetrahedral values. The chair conformation of cyclohexane has no strain energy— the bond angles are ideal, and the conformation about each C–C bond is staggered. That helps to explain why cyclohexane rings are found so prevalently: they are unstrained and very stable!

Confirmation of Sachse's chair model took about 25 years, when Ernst Mohr, a German chemist, published the x-ray crystal structure of diamond, with a portion of this repeating structure shown in Figure 11.6. This image of diamond shows each carbon atom bonded to four other carbon atoms in a tetrahedral arrangement. Six carbon atoms form a ring, and these rings are all

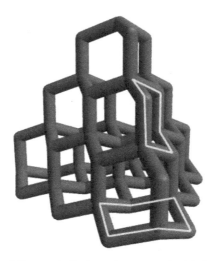

**Figure 11.6.** Structure of diamond. Chair conformation highlighted in white.

in the shape of the chair conformation. I have highlighted two of these chair conformations in white.

Let me take a short digression to discuss x-ray crystal structures. I suspect that what everyone who studies chemistry at any point wants, whether it's in grade school, high school, or college, is a picture of a molecule, giving them some way to "see" the molecule, its constituent atoms, in some tangible way. We know that molecules are really small, but using some type of microscope, why can't we take a picture of them? Well, x-ray crystallography is a way to do just that!

Imagine grabbing a few pebbles and throwing them into a pond. Each pebble will create a concentric circle of waves that spread outward from where they land in the water. Now take a picture of the pond a few seconds after you tossed the pebbles; that image might show the tops of the waves as in Figure 11.7. By examining the wave tops, their crests, we can infer where the pebbles entered the water. The point here is that the image helps us identify the location of objects that we haven't directly observed; we have not used a picture of the pebbles as they splashed into the water.

Let's move on to a related property of waves. When a flat (plane) wave passes through a slit, it will *diffract*; that is, the wave passes through the slit now, making a semicircular form. It's as if the wave in the middle of the slit moves through unimpeded while the wave at the edge of the slit gets held back a bit. An interesting phenomenon occurs when a wave passes through multiple slits. At each slit, the wave propagates as increasingly large semicircles. The semicircular crests will then overlap to build even taller crests, or what we

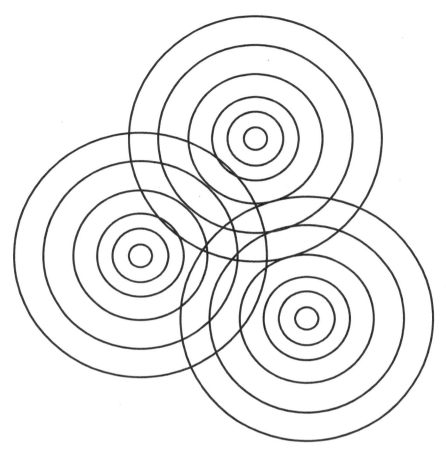

**Figure 11.7.** Schematic of the spreading ripples from pebbles dropped in a pond.

call *constructive interference*. There will also be regions where the crest from one semicircular wave will hit the trough of a neighboring wave, and they will cancel each other out, or what may be called *destructive interference*. This pattern is shown in Figure 11.8, showing a plane wave passing through two slits and setting up a pattern of constructive interference.

Now imagine that you can only observe where the constructive interference takes place. You can't see the slits. You can't see the destructive interference because the waves cancel out there. That doesn't seem like much information, but it is actually enough to determine where the slits are in relation to each other. I won't go through the math here, of course, but trust me.

What does any of this have to do with obtaining the structure of a molecule? X-rays are a form of light, and they behave as waves, just like the waves in a pond or on the ocean. When x-rays pass through a molecule, they will

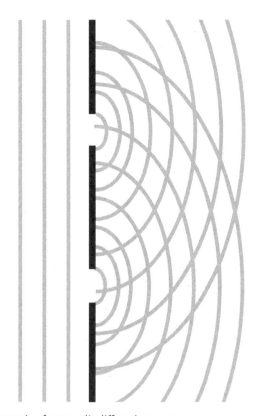

**Figure 11.8.** Schematic of a two-slit diffraction pattern.

diffract off the atoms, just like passing through slits. A crystal is a solid where the molecules pack together in repeating patterns, setting up periodic "slits" for the x-rays to diffract through. Detecting the locations and intensities of the constructive interference provides the raw data, and with some math and the help of modern computers, this data can be converted into the locations of the atoms. And from there, we can draw pictures representing the atoms in a molecule, just like the picture of diamond in Figure 11.6.

As you might imagine, being able to "see" molecules, to get a real sense of their geometries and shapes, provides tremendous information for chemists when they design molecules with specific purposes. For example, most drugs work by binding, or attaching, to a protein, thereby activating or deactivating the work of that protein. Drug design is greatly accelerated if we know the shape of the protein, and especially the binding site of that protein.

Since x-ray crystallography provides such important data, it is not surprising that many Nobel Prizes have recognized scientific developments in this field. The German physicist Max von Laue won the 1914 Nobel Prize in Physics for the discovery of x-ray diffraction. The following year, the father and son team, William Henry and William Lawrence Bragg, were awarded the physics prize for their work in solving the first set of molecular structures. Jerome Karle and Herbert Hauptmann were awarded the 1985 Nobel Prize in Chemistry for their development of mathematical methods that greatly facilitated solving crystal structures. Max Perutz and John Kendrew shared the 1961 Nobel in Chemistry for determining the structures of hemoglobin and myoglobin, respectively; these are molecules critical to the transport of oxygen in blood. That same year the Nobel Prize in Medicine was awarded for perhaps the most famous x-ray structure ever: to James Watson, Francis Crick, and Maurice Wilkins for their determination of the structure of DNA. It is one of the tragedies in science that Rosalind Franklin, the person who *actually conducted the x-ray diffraction study* of DNA, passed away prior to the awarding of this Nobel Prize. A number of other Nobel Prizes have been awarded for the determination of other essential components of the life process, including the structure and function of the ribosome and membrane channels.

All right, that was a long digression. Now back to the chair structure of cyclohexane. Look closely at the chair model in Figure 11.5. You will notice three sticks (three hydrogens) pointing straight up and another three pointing straight down. The other six sticks (or hydrogen atoms) are pointing outward to the side. We refer to the hydrogens pointing up or down as *axial* and the ones pointing to the side as *equatorial*, since they mimic being along the equator of the earth. The six axial hydrogens are identical to each other, since simple rotation of the entire molecule will interconvert them. The same is true of the six equatorial hydrogens; they too are identical to each other. However, the axial hydrogens are different from the equatorial hydrogens. Note that the three axial hydrogens on the top are near each other, just as the three axial hydrogens on the bottom. But the equatorial hydrogens do not get near each other. Thus, the hydrogens in the chair conformation divide into two sets.

Interestingly, room temperature observation of cyclohexane reveals that all twelve of the hydrogen atoms are identical to each other. I'll get to the experimental evidence in a bit, but first I will explain how to make all twelve hydrogens identical.

Remember that rotation about a C–C single bond is very easy, with a small activation barrier. Let's do the following rotations, diagrammed in Figure 11.9. Imagine grabbing carbon #1 and pulling it upward. That will entail rotation about the C1–C2, C2–C3, C1–C6, and C5–C6 bonds, leading to the middle structure of Figure 11.9. We call that structure a *boat* conformation. Next,

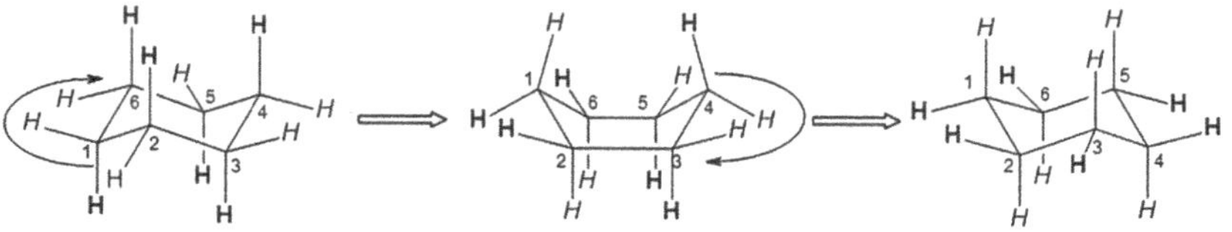

**Figure 11.9.** Steps to perform a cyclohexane ring flip.

grab the C4 carbon and rotate it downward. This involves rotations about the C4–C3, C3–C2, C4–C5, and C5–C6 bonds, leading to the chair conformation on the right.

It may seem that this has been a pointless exercise. All we've done is rotate a chair conformation into a chair conformation. However, look very closely at the hydrogens. In the starting conformation, I have labeled all of the axial hydrogens in bold and all of the equatorial hydrogens in italics. (Remember that in principle we can really do that—we could have all the axial positions as hydrogen atoms and all of the equatorial positions as deuterium atoms.) When the two sets of rotations are done, in the final chair conformation, the bold hydrogens are now equatorial and the italic hydrogens are axial. The hydrogens have flipped positions: what was axial has become equatorial, and what was equatorial has become axial. This conformation change, which we call a *ring flip*, exchanges the axial and equatorial positions. This ring flip and its exchange of the axial and equatorial positions can also be seen in Figure 11.5. The axial (white) hydrogens in the chair conformation to the left end up in the equatorial positions in the ring-flipped chair conformation on the right side.

To understand the experimental observation of the conformation of cyclohexane, let's consider photographing a waterfall. In most pictures of waterfalls, the falling water appears as a blur. That's because during the time that the picture is being taken, the water moves significantly and what the camera sees is the position of the water blurred over space. You have undoubtedly seen photographs where the water appears as frozen droplets fixed in space. It's not that the water just stopped midflight and waited for the photographer to snap the picture. Rather, the photographer used a very short shutter speed, meaning that that the film was exposed for a very short time, short enough that during the time the film was exposed, the falling water barely moved.

Cyclohexane will undergo rapid ring flipping at room temperature. Any "picture" taken of cyclohexane will show a blurred image of the hydrogens, meaning an average of being in the axial and equatorial positions. At low temperature, the ring flipping can be slowed, and at sufficiently low temperatures, there will not be enough energy available to the molecule to overcome the barrier to ring flipping. This will freeze every cyclohexane molecule into a fixed chair conformation. In this frozen state, a "picture" would identify two distinct sets of hydrogens: the axial and the equatorial sets.

Now the "picture" we can take to see the ring flip of cyclohexane is nuclear magnetic resonance (NMR), a method related to magnetic resonance imaging (MRI). Each hydrogen nucleus has a magnetic dipole, a north pole and a south pole. We call this magnetic dipole associated with a nucleus *spin*. When a compound is placed inside a magnetic field, these dipoles, or spins, line up with the magnet. When the compound is blasted with radio waves, a specific frequency (think of this as tuning in a specific radio station) will cause the spin to flip. The NMR experiment detects these spin flips. At room temperature, we see a single signal in the NMR of cyclohexane, telling us that all of the hydrogens are identical; what we observe here is an average (a blurring) of the axial and equatorial positions. At low temperature, the ring flip stops, and we see two signals in the NMR, one for the axial hydrogens and one for the equatorial hydrogens.

Monosubstituted cyclohexanes will have two observable chair conformations, one where the substituent is in the axial position and one where it is in the equatorial position, such as in methylcyclohexane, shown in Figure 11.10. The double-headed arrow in the reaction indicates that the reaction is constantly going forward and backward. In this case, the $CH_3$ groups are not equivalent in the two molecules. They can be distinguished in the NMR as two separate signals, with a greater proportion of the molecules having the methyl group in the equatorial position. That the equatorial positon is more favorable than the axial is understood in terms of *steric interactions*, where groups get close and bump up into each other. The axial methyl group interferes in the space of the two axial hydrogens, as seen in the conformation on the left side of Figure 11.10. In general, a more favorable conformation has as many equatorial substituents as possible.

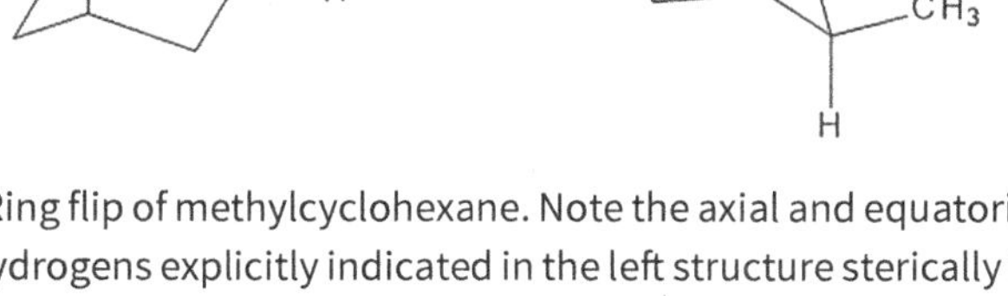

**Figure 11.10.** Ring flip of methylcyclohexane. Note the axial and equatorial positions. The two axial hydrogens explicitly indicated in the left structure sterically interfere with the methyl group.

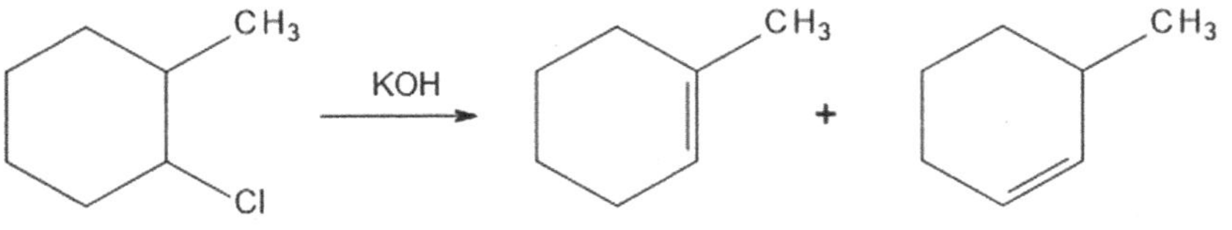

**Figure 11.11.** Elimination reaction of 1-chloro-2-methylcyclohexane **1** with KOH.

Returning now to the anti-periplanar elimination of the E2 reaction, let's consider what this means for reactions involving substituted cyclohexanes. Looking back at Figure 11.3, how can the hydrogen that will be removed by the base and the leaving group be positioned in an anti-periplanar way? The only option is for them to be in the axial position on adjacent carbon atoms. Any other arrangement—both equatorial or one axial one equatorial—will not get them into the same plane, let alone on opposite sides.

This provides an excellent opportunity to again test the anti-periplanar hypothesis. The reaction of 1-chloro-2-methylcyclohexane **1** with base (KOH) provides an excellent example (Figure 11.11). Elimination of HCl from **1** will in principle create two alkenes, the more substituted alkene **2** and the less substituted alkene **3**. We might anticipate that the more substituted alkene **2** will be the major product, since we are using a very small base and we should make the thermodynamic product.

There are two isomers of **1**: the *cis* isomer, which has the two substituents on the same face of the ring, and the *trans* isomer, which has the two substituents on the opposite faces of the ring (Figure 11.12). Elimination from the *cis* isomer produces the more stable alkene **2** as the major product, with only a small amount of **3** produced. However, the *trans* isomer reacts with the base to produce mostly the less substituted alkene **3**, with only a trace amount of **2** formed. How do we rationalize these results?

We might at first just consider the reaction of *cis*-**1** as representing a case of thermodynamic control, where the more stable product is predominantly produced. However, that provides no guidance toward understanding the results of the reaction of *trans*-**1**. Instead, let's consider that an E$_2$ reaction will proceed by anti-periplanar elimination. That requires that the eliminating groups, H and Cl, be in axial positions. I have drawn the chair conformations of *cis*-**1** and *trans*-**1** in Figure 11.13. In both cases, I have drawn the more stable chair conformation on the left. For *cis*-**1**, the chloro substituent has a slightly greater propensity to be in the equatorial position than does the methyl group, but the two conformations are close in energy. It's no competition with *trans*-**1**, as the left conformation has

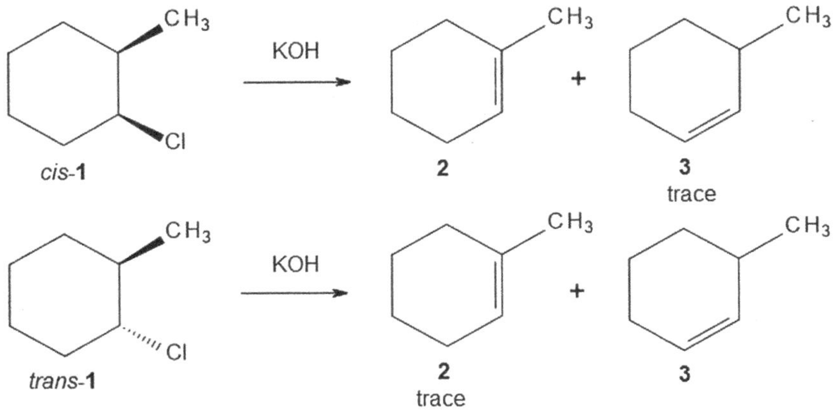

**Figure 11.12.** Reaction of *cis*-1 and *trans*-1 with KOH.

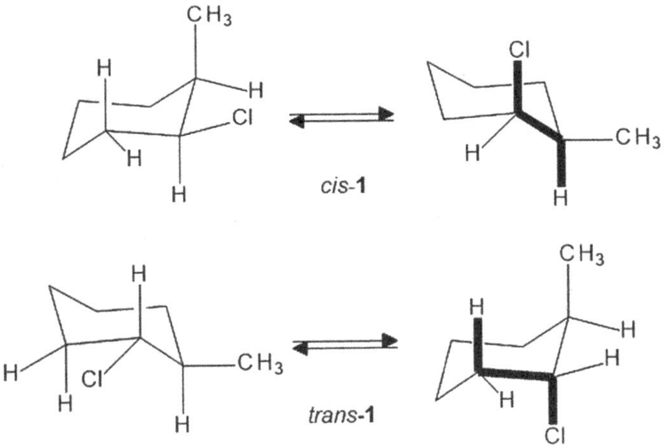

**Figure 11.13.** Ring flip chair conformations of *cis*-1 and *trans*-1. The anti-periplanar elimination is highlighted in bold.

both substituents in equatorial positions, while the right conformation has them both in the less favorable axial positions.

Note that in the left, more stable, chair conformation for both the *cis* and *trans* isomers, the chloro group is in an equatorial position. No anti-periplanar elimination is therefore possible. After ring flip, however, the chloro group is in the axial position. For *cis*-1, there is a neighboring hydrogen in the axial position, on the carbon that also has the methyl group. Anti-periplanar elimination, highlighted by the relationship shown in bold in Figure 11.13, will lead to **2**, the more stable alkene. That's the major product observed. For the *trans*

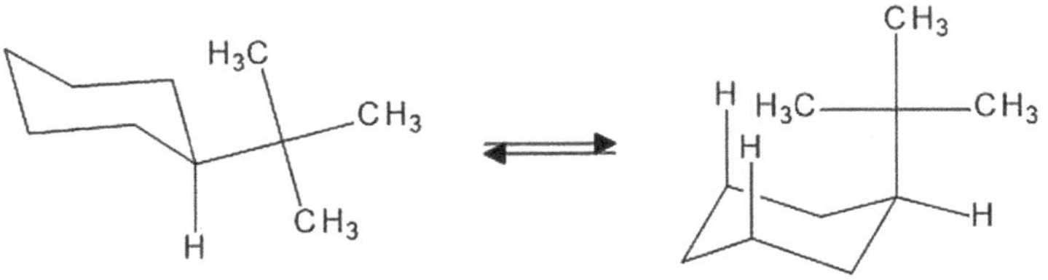

**Figure 11.14.** Chair conformations of *t*-butylcyclohexane. The conformation on the right, with the axial t-butyl group, is essentially precluded by steric interactions.

isomer, the only neighboring hydrogen in an axial position is on the carbon to the left. This anti-periplanar arrangement, highlighted in bold in Figure 11.13, leads to the less substituted alkene **3**. Again, that's in agreement with the experimental observation. This is a particularly strong argument in support of the anti-periplanar mechanism; the observed product from elimination of *trans*-**1** is rationalized as coming about via the anti-periplanar route, despite having to proceed through the much less stable chair conformation *and* it produces the less stable alkene!

I close this chapter with one last example, this one involving a *t*-butyl substituted cyclohexane. The *t*-butyl substituent is quite bulky, as shown in Figure 11.14. This bulk significantly destabilizes the conformation where it is in the axial position, to the extent that we detect virtually none of this axial conformation. We consider the *t*-butyl group to lock the conformation so that it remains in the equatorial position.

Let's examine the implications of that locked conformation in the reaction of *cis*-**4** and *trans*-**4** with the base potassium hydroxide (KOH). Elimination from the *cis* isomer produces the alkene **5**. However, the major product from the reaction with *trans*-**5** is the substitution product **6**, while the elimination product **5** is produced very slowly, indeed, much more slowly than in the reaction of the *cis* isomer (Figure 11.15).

As you might expect, understanding these results requires examination of the conformations of the two isomers of **4**. Since the *t*-butyl group locks the chair and prohibits ring flipping, we need only look at one chair conformation for the *cis* isomer and one chair conformation for the *trans* isomer, each with the *t*-butyl group in the equatorial position. These are drawn in Figure 11.16. The chair conformation of *cis*-**4** places the chloro group in an axial position, set up for anti-periplanar elimination (shown in bold) that leads to **5**. However, the chloro group is in an equatorial position in *trans*-**4** and has no

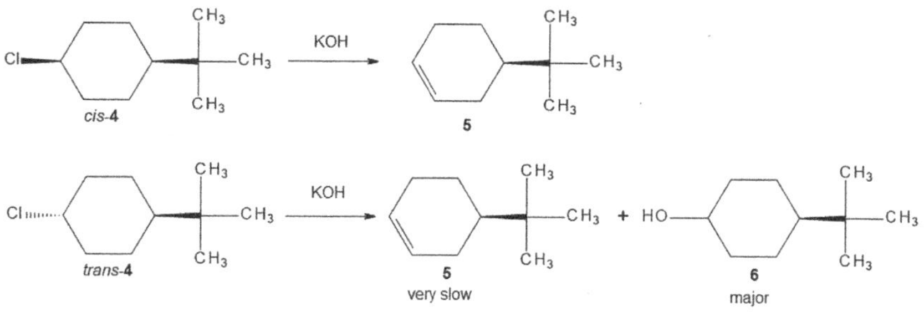

**Figure 11.15.** Elimination reaction of *cis*-4 and *trans*-4 with KOH.

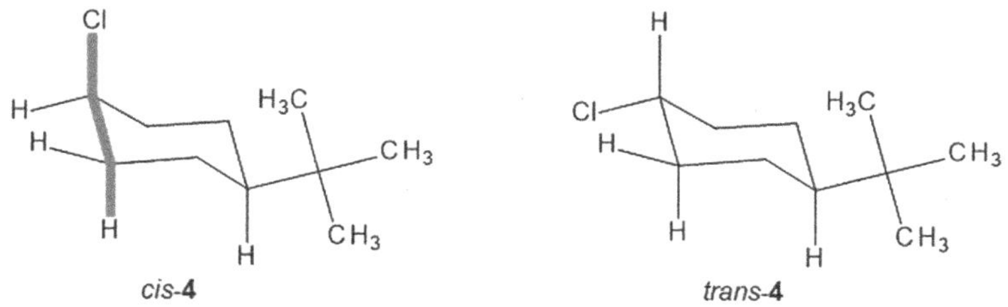

**Figure 11.16.** Chair conformation of *cis*-4 and *trans*-4. The anti-periplanar pathway is highlighted in bold.

anti-periplanar route, precluding the E$_2$ mechanism. Instead, what is taking place is first the loss of chloride (Cl$^-$), to create the secondary carbocation. Next, there is a competition between nucleophilic attack by hydroxide (S$_N$1), leading to **6** or a base attack by hydroxide (E$_1$) leading to **5**. The fact that the elimination is so much slower for *trans*-4 than for *cis*-4 suggests different mechanisms—E$_1$ for the former and E$_2$ for the latter.

These elimination reactions with substituted cyclohexanes provide some insight into the reaction pathway, indicating how the reacting species come together and their orientation and movement as the reaction progresses. Keep in mind how radical that intellectual outcome really is. In the classical world, we would just film the process and then run it in slow motion, following, say, a perfectly thrown spiral launched by the quarterback, progressing in a graceful arc into the outstretched hands of the receiver racing down the sidelines. Imagine the film showing in slow motion the carom of the cue ball striking the eight ball that then bounces off a rail and into the side pocket.

Quantum mechanics provides no knowledge of where a particle is and where it is going. To acquire some knowledge we must instead rely on indirect evidence of a reaction path, like what we obtain from these studies of the reactions of isomeric conformations or from the elegantly designed endocyclic restriction test discussed in Chapter 10. Quantum mechanics imposes some hard challenges for us, but creative thought and execution of clever experiments can allow us to obtain important understanding of the details of a reaction.

# 12

# The Truth about Substitution and Elimination Reactions

When I arrived at the University of California-Berkeley to start my PhD studies, I felt very prepared. The chemistry curriculum I studied at the University of Illinois at Urbana had been rigorous and comprehensive. I thought I was especially well prepared for organic chemistry, having taken two additional organic chemistry classes beyond the introductory year. What was in store for me was a major surprise. Organic chemistry was *not* what I had been taught as an undergraduate!

I probably should have expected nature to be more complex than what is presented in an undergraduate class and textbook. I was just too naïve at that time to think through the process of making science understandable to novices and to understand what sort of simplifications and selective omissions were needed to create a class digestible to a brand-new student.

One particular class experience should have given me some pause to reflect on how to deliver ongoing science to an undergraduate class. I was nearing the end of the fall semester of my senior year and had enrolled in the subatomic physics class. I was extremely excited to take this class; from the time I entered college, I had planned my course sequence in order to take it. Even though I was a chemistry major and had decided to focus on organic chemistry, I was fascinated by subatomic physics. I was intrigued by how physicists had constructed the hierarchy of subatomic particles. How did they discover these particles? How did they obtain information on their properties? What rules governed their behavior? What new particles might still be awaiting discovery? It all seemed so otherworldly, representing such an amazing intellectual achievement, and I wanted in.

One of the subatomic particles we discussed in the class was the neutrino. It comes in three different "flavors" (you've got to love physicists' knack for nomenclature). The neutrino was posited to be massless, just like a photon, the quantum manifestation of light. As a massless particle, the neutrino would stay as one type of flavor. Before I took the subatomic physics course, there were some indications that his lack of interconversion of neutrino type, which is better called *oscillation*, might be false. But our professor led us through the

*Thinking Like a Physical Organic Chemist.* Steven M. Bachrach, Oxford University Press. © Oxford University Press 2023.
DOI: 10.1093/oso/9780197640371.003.0012

material guided by the current accepted belief that the neutrino was massless. About a week before the end of the semester, a group of physicists reported some preliminary work suggesting that the neutrino did have mass. Even though the proposed quantity of mass was very small, *any* amount of mass meant that neutrino oscillation would occur. That implied that a whole lot of what we had just learned that semester was, well, maybe not wrong, but in need of serious revision.

I remember being dismayed and disheartened at the time, thinking that much of the effort I had expended to learn the materials was for naught, that the world had changed. Instead, I should have been overjoyed at witnessing science in action! Science is exciting because new experiments continually lead us to reevaluate our current ideas and create new models. They challenge us to devise additional experiments, test our assumptions, and develop new knowledge. If every experiment worked the way you anticipated, that would be very boring, leading to nothing novel. It's the unexpected result, the experiment that counters our conventional thought, that makes a difference and leads to paradigm changes.

It turns out that the particular neutrino experiment was actually inconclusive, but it did inspire a number of physicists to perform more careful, more sensitive experiments. It wasn't until about seventeen years later that the definitive experiments were performed confirming neutrino oscillation, and then about another eighteen years until the Nobel Prize in Physics was awarded to the leaders of the experiments. Even today, we still do not know the actual mass of the neutrino in their three flavors; we only know the difference of the squares of their masses.

I don't remember what I expected to learn from my first course in organic chemistry as a graduate student, but I certainly didn't expect to be slapped in the face. That's what it felt like to learn that organic chemistry was not black and white, but infinite shades of gray.

Let's walk through a few of the experiments that shattered my primitive view of organic chemistry. We start with the notion that a tertiary system will undergo the $S_N1$ reaction only, that the $S_N2$ reaction is prohibited because the backside is blocked off. The reaction of *t*-butylchloride 1 with ethanol is typically presented as the paragon of the $S_N1$ reaction (Figure 12.1).

The German-American chemist Paul von Rague Schleyer worried that the backside of 1 might not be totally inaccessible to a nucleophile. He cleverly devised the tertiary system adamantylchloride 2, which might look somewhat familiar as the building block of diamond. The backside of the carbon bonded to chlorine is blocked here by the entirety of the cage structure. Truly,

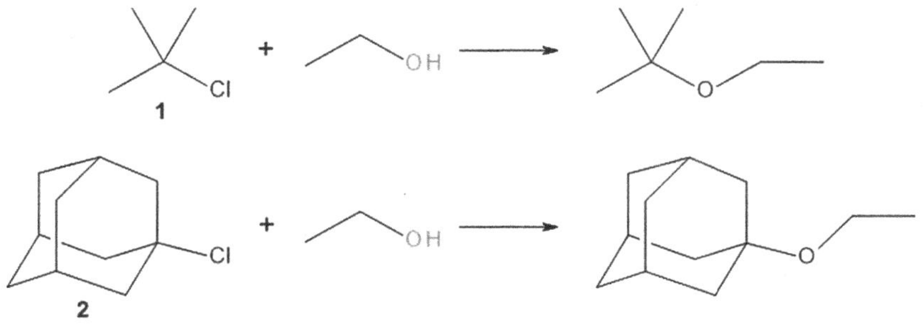

**Figure 12.1.** Solvolysis reactions of *t*-butylchloride and adamantylchloride.

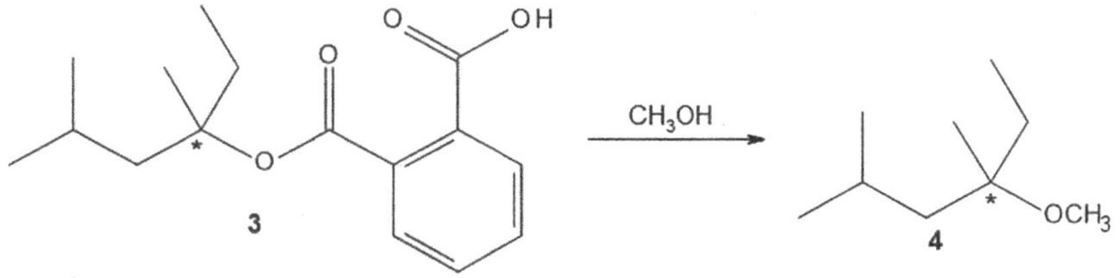

**Figure 12.2.** Solvolysis of **3**. The chiral center marked by an asterisk (*).

no nucleophile can make its way through the cage to attack that carbon and push out the chlorine.

Schleyer compared the rates of reaction of **1** and **2** with different nucleophiles. With a very weak nucleophile, the rates are very close. However, with the stronger nucleophile ethanol (as shown in Figure 12.1), the reaction of **1** is 3000 times faster than with **2**. With a weak nucleophile, the two substrates behave similarly, but with an active nucleophile, one that wants in on the action, it appears that the nucleophile does help push out the leaving group of **1**. The nucleophile can't assist in the reaction of **2** because it can't get to the backside. That's compelling evidence that a nucleophile is assisting the reaction through backside attack in **1**, and that's a key element of the $S_N2$ mechanism.

The second experiment is representative of a number of studies. The $S_N1$ mechanism posits a planar carbocation intermediate. The nucleophile should then attack equally well from either side of that molecular plane. If the starting substrate is chiral, this type of attack should lead to a racemic product, equal amounts of the left-handed and right-handed molecules. The American chemist William von Eggers Doering performed the reaction shown in Figure 12.2. The chiral compound **3** reacts with methanol, $CH_3OH$, to produce **4**. When the nucleophile is also the solvent, we call this a *solvolysis reaction*. The

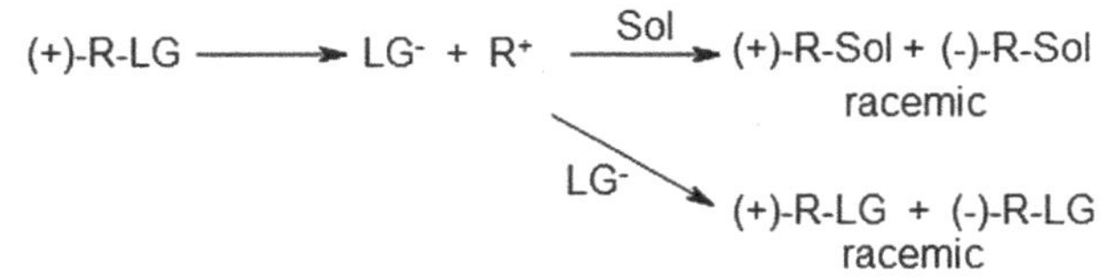

$$(+)\text{-R-LG} \longrightarrow \text{LG}^- + \text{R}^+ \xrightarrow{\text{Sol}} (+)\text{-R-Sol} + (-)\text{-R-Sol}$$
$$\text{racemic}$$

$$\text{LG}^- \searrow$$
$$(+)\text{-R-LG} + (-)\text{-R-LG}$$
$$\text{racemic}$$

**Figure 12.3.** Schematic of Winstein experiment of the solvolysis of a chiral compound.

chiral center is identified with an asterisk. Note that the chiral center is tertiary, so this should proceed via the $S_N1$ mechanism and lead to a racemic product. However, the resulting product comes from 72% inversion at the chiral center and 28% retention, a far cry from the expected 50:50 ratio. Could this mean that a significant percentage of the reaction takes place via the $S_N2$ pathway to yield that large amount of inverted product? That seems unlikely.

Saul Winstein, a preeminent physical organic chemist from the United States, conducted quite a few experiments to clarify the picture of nucleophilic substitution. One of his important experiments is schematically diagrammed in Figure 12.3. It begins with a chiral substrate, which we will designate as (+)-R-LG. The plus sign indicates that it is the enantiomer of R-LG that will rotate polarized light clockwise. Let's assume that R-LG will undergo solvolysis following the $S_N1$ mechanism. The first step is the loss of the leaving group to create the planar carbocation $R^+$. There are two possible paths going forward. First, the solvent could attack the carbocation equally from either side of the planar $R^+$, yielding a racemic solvolysis product (+)-R-Sol and (–)-R-Sol. The other possibility is for the leaving group to act as a competing nucleophile and to reattach. It, too, should attack from either face, leading to both (+)-R-LG and (–)-R-LG in equal amounts.

Over time, we would expect to see growing amounts of racemic R-Sol and the racemization of the starting material R-LG. We might also expect the rate of formation of product R-Sol to be faster than the rate of racemization. This is because so much more solvent than $LG^-$ is present, such that the odds of the $R^+$ cation finding the $LG^-$ before running into and reacting with a solvent molecule are very small. That's the prediction: the rate of solvolysis should be much faster that the rate of racemization of the reactant.

Shown in Figure 12.4 is the experiment that Winstein actually conducted, the solvolysis of chiral **5** with acetic acid to give **6**. He determined that the rate of racemization of **5** was about *30 times faster* than the rate of solvolysis, the production of **6**. This is opposite to our expectation according to our $S_N1$ model.

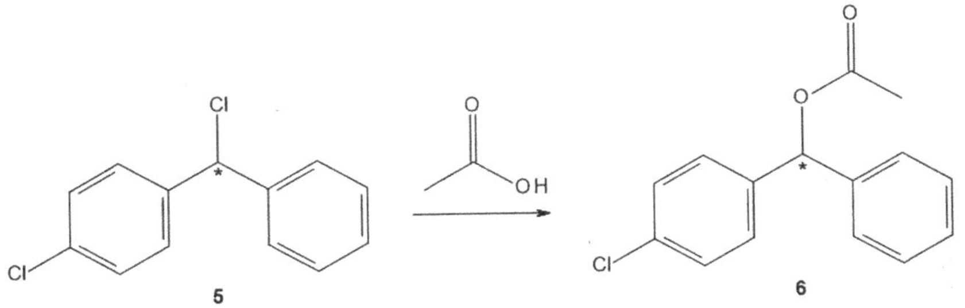

**Figure 12.4.** Solvolysis of **5**. The chiral center is marked by an asterisk (*).

In order to set up the next experiment, let me present an analogy. Imagine a group of 100 ten-year old girls aspiring to be baton twirlers gathered on a field. Each girl has a baton with one end colored green and the other end colored blue. Each twirler begins by holding the green end. Now a practiced baton twirler would likely be able to flip the baton in the air and catch the green end nearly every single time. But these are novice twirlers. Most of them will catch the green end of the baton, but a few will end up catching the blue end.

Every time they throw the baton into the air, the twirlers will spin around. They all start by throwing the baton with their right hand. Most will catch it back with their right hand. But some will spin too little or too much and have to catch the baton with their left hand. Lastly, if the twirler drops the baton, they leave the field.

At the start, there are 100 girls on the field, each holding the green end of the baton in their right hand. The instructor blows the whistle, and the twirlers start doing their thing. Batons are thrown into the air, the girls are spinning around, batons are caught. Occasionally, the baton is dropped and the girl leaves the field. They twirl for 30 seconds, and the whistle is blown and everyone stops. We go around and count how many batons are held at the green end and at the blue end; we count how many batons are in the left hand and in the right hand; and we count how many girls have left the field. The whistle is blown and the remaining girls continue their baton routine for 30 seconds. They stop and we repeat our count. This process is repeated a few times.

At the end we obtain three different rates. First, we compute the rate of scrambling of the green or blue end being held. Second, we compute the rate of racemization by looking at which hand holds the baton. Third, we compute the rate at which the girls leave the field due to dropping the baton.

Now let's look at an exquisitely designed nucleophilic substitution experiment that provides three analogous rates to those of our baton twirlers. Harlan Goering constructed **7** (Figure 12.5), which has much in common with the

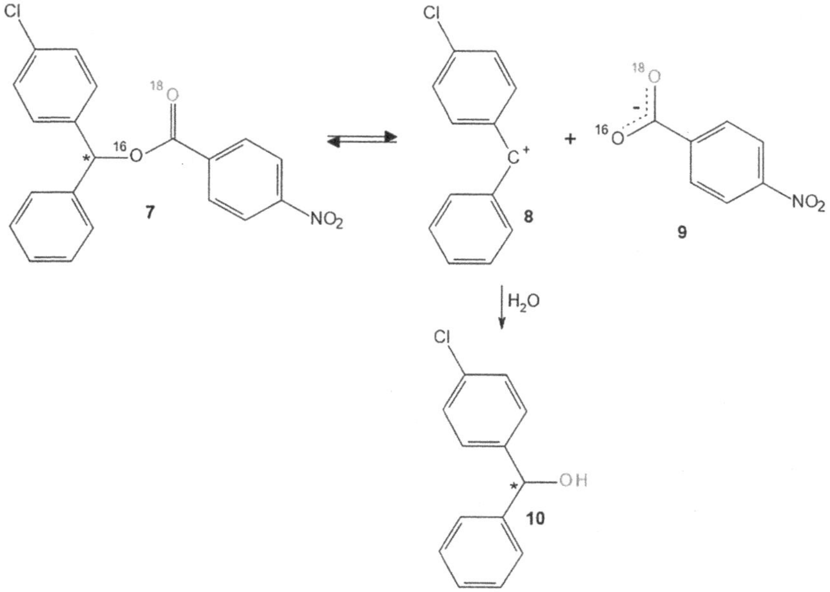

**Figure 12.5.** Solvolysis of **7**. The chiral center is marked by an asterisk (*).

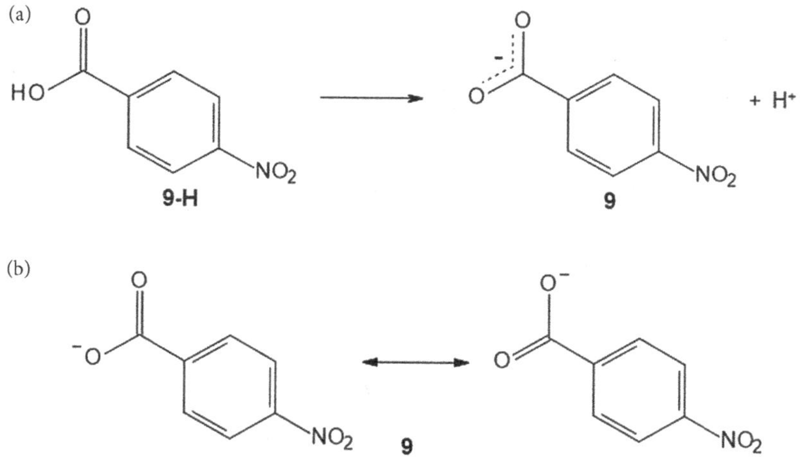

**Figure 12.6.** (a) Acid reaction of **9-H**. (b) Resonance structures of **9**.

reactant in our previous example, compound **5**. Both have a tertiary chiral center but differ in their leaving group.

The leaving group in **7** is p-nitrobenzoate **9**, which is the anion created when p-nitrobenzoic acid acts as an acid and loses H$^+$ (Figure 12.6a). The key feature of **9** is that normally the two oxygen atoms would be indistinguishable.

Multiple different types of experiments of many molecules similar to **9** show this equivalence; perhaps the best evidence is that the two C–O bond lengths are identical. How do we rationalize that equivalence? Let's look at the structure on the left in Figure 12.6b. The double bond is drawn to the oxygen at the top, and the negative charge resides on the oxygen to the left. But that's an arbitrary choice—it's just as proper to draw the double bond to the oxygen on the left and to have the negative charge on the oxygen at the top, that is, the structure on the right in Figure 12.6b. Which is correct? Each individually is a valid way of distributing the electrons, but it fails to match the experiments; each structure predicts one short C–O bond and one long C–O bond, not two equivalent distances. The American chemist and twice Nobel Laureate recipient Linus Pauling developed the notion of resonance, and argued that the proper description of this molecule is a 50:50 combination of these two structures, that the carbon and oxygen are joined by a bond of order 1.5, and that each oxygen carries are charge of –0.5. That's what is meant by the dotted lines between the carbon and oxygen atoms in the depiction of **9** in Figures 12.5 and 12.6a.

What is so clever in the design of **7** is that Goering makes the two oxygen atoms distinguishable through the use of isotopic substitution, yet they remain *chemically identical*. Just as I discussed earlier about the isotopes of hydrogen, oxygen exists as a few isotopes. The most common isotope is $^{16}O$, which has eight protons and eight neutrons. A very small amount of $^{18}O$ exists on earth, and it has eight protons (that's what makes it oxygen!) and ten neutrons. Goering was able to synthesize **7** with the $^{18}O$ in the position of the double bond. Once the leaving group takes off, resonance in the resulting anion makes the two oxygen atoms chemically identical, meaning they will have identical reactions. The isotopic label allows us to track each atom separately to identify what happens to each atom. So, the leaving group can exit from **7** and form the anion **9** and the cation **8**. These could reattach, but now the reattachment might return through the $^{16}O$ atom or through the $^{18}O$ atom. This is depicted in Figure 12.7. Since we can track these two different isotopes, we can measure the rate of this exchange. This is akin to measuring

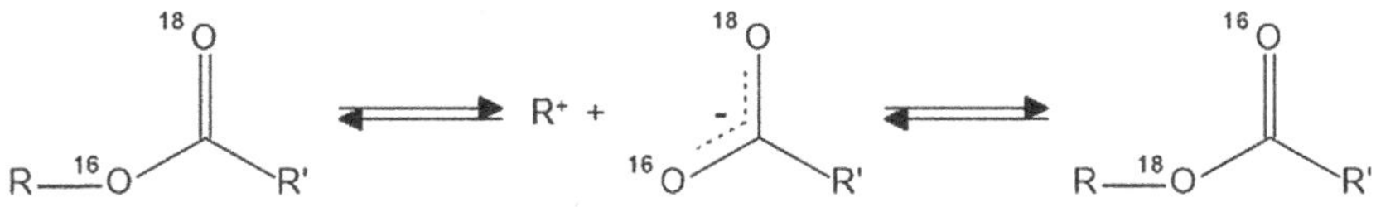

**Figure 12.7.** Scrambling of the oxygen atoms in the Goering experiment.

$$(+)\text{-R-O}_2\text{CR'} \rightleftharpoons \text{R}^+ + {}^-\text{O}_2\text{CR'} \rightleftharpoons (-)\text{-R-O}_2\text{CR'}$$

**Figure 12.8.** Racemization of the chiral center in the Goering experiment.

$$\text{R-O}_2\text{CR'} + \text{H}_2\text{O} \longrightarrow \text{R-OH}$$

**Figure 12.9.** Solvolysis reaction in the Goering experiment.

the rate at which the baton twirlers catch the blue end instead of the green end of the baton.

Next to consider is that anion **9** and cation **8** can recombine with retention of the initial stereochemistry or with inversion, as shown in Figure 12.8. We can measure the resulting rate of racemization of **7**. This rate is akin to the rate at which the baton twirlers, who all start with the baton in the right hand, end up catching the baton in their left hand.

The last part of Goering's experiment was to measure the rate of overall reaction, the rate of solvolysis. This reaction is depicted in Figure 12.9. It is analogous to the rate at which the baton twirlers drop their batons and leave the field.

What would we expect of these three rates if the reaction takes place according to the $S_N1$ reaction discussed in the earlier chapters? The first step of this mechanism is for **7** to cleave into the cation **8** and anion **9**. The cation can then be attacked by the nucleophile, which in this case is the solvent, water. The cation could recombine with the leaving group. However, since there is so much more solvent around than the anion **9**, the odds are hugely in favor of the solvolysis reaction, and very little recombination should take place. That suggests that the rate of equilibration of the two oxygens and the rate of racemization should be very small, almost negligible, and the rate of solvolysis should be dramatically larger than the other two rates.

What Goering actually observed is that the rate of solvation and the rate of racemization are about the same and that the rate of equilibration of the two oxygens is about 1.5 times faster. Clearly, this observation is not compatible with the simple $S_N1$ model presented earlier, as are the other results presented in this chapter.

So what is really going on? The $S_N1$ and $S_N2$ mechanisms I described in earlier chapters are *exactly what are presented in almost every introductory organic chemistry textbook*. These mechanisms can't be totally wrong, or a whole lot of organic chemistry professors are committing fraud! In fact, the $S_N1$ and $S_N2$ mechanisms are limiting cases that are observed in some specialized circumstances. The truth is more complicated.

We need to modify the $S_N1$ mechanism by taking a closer look at what happens after the C–LG bond breaks, creating two ions. These two ions, the carbocation and LG⁻, do not simply swim away from each other, to be *free ions* moving independently within the solvent. Rather, there are a series of intervening *ion pairs*, where the cation and anion stay near each other.

The first ion pair is called the *contact ion pair*, where the two ions stay adjacent to each other, surrounded by solvent. This is the first structure formed upon cleavage of the bond. It is shown as the first ion pair in the reaction displayed in Figure 12.10.

Three things can happen to the contact ion pair (CIP). First, it can recombine (or go backward in the reaction of Figure 12.10). Since the two ions are right next to each other, the leaving group will attack from the same side that it exited, leading to retention of the stereochemistry of the starting material. However, in the case of 7, though anion 9 stays close to cation 8, the reattachment could easily go to either oxygen, leading to scrambling of the oxygens. Second, the nucleophile might attack the carbocation. If this were to happen, the nucleophile would only be able to attack from the backside; the frontside is blocked by the leaving group that remains adjacent. Since in the experiments presented here, the nucleophile is the solvent, this attack is certainly possible since the solvent surrounds the ion pair. So, any nucleophilic attack that happens with the contact ion pair would lead to inverted stereochemistry in the product. In other words, one can get inversion even though it's not an $S_N2$ process! Third, the contact ion pair can proceed along the pathway shown in Figure 12.10, forming a second type of ion pair.

The *solvent separated ion pair* still has the two ions near each other but with one or two solvent molecules between them. This is pictured with the two vertical lines between the ions in Figure 12.10. The cation and anion are still close enough that their electrical attraction can maintain the pair, but with an intervening solvent molecule or two. While the two ions are still near each other, their separation means that each ion can move more than what was possible within the contact ion pair. In particular, the carbocation, which is planar,

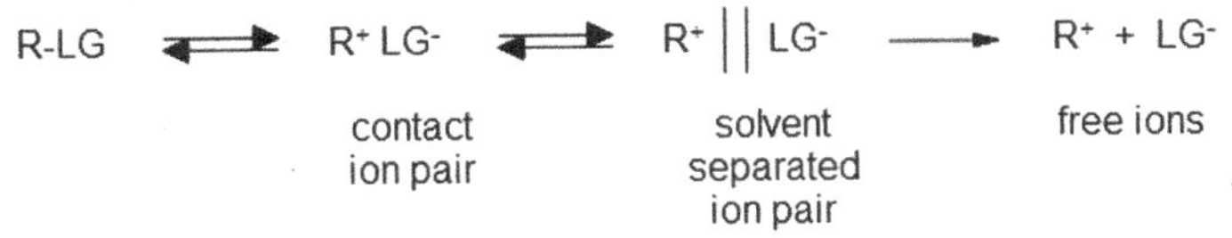

R-LG  ⇌  R⁺LG⁻  ⇌  R⁺ ‖ LG⁻  ⟶  R⁺ + LG⁻

|  | contact<br>ion pair | solvent<br>separated<br>ion pair | free ions |

Figure 12.10. Ion pair interconversions.

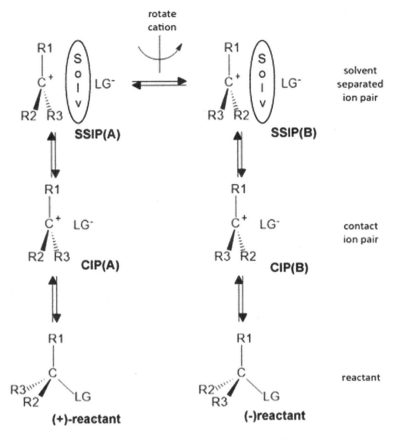

**Figure 12.11.** Mechanism for racemization of reactant in a solvolysis reaction.

might spin around, such that both faces of the plane are exposed to the neighboring anion. This is pictorially represented at the top of Figure 12.11.

Let's now consider what can happen to the solvent-separated ion pair (SSIP). Over time, the original **SSIP(A)** will convert to **SSIP(B)**. First, each of them can then revert back to their contact ion pair form, **CIP(A)** and **CIP(B)**, respectively. These contact ion pairs might revert back to reactant, leading to racemization!

Second, the two solvent separate ion pairs might be attacked by the nucleophile. Again, since in solvolysis reactions, the solvent is the nucleophile, the nucleophile is nearby and ready to attack. In the SSIPs, the solvent might attack from the backside or even from the frontside: a solvent molecule resides between the two ions, that is, from the front side. This would lead to both retention and inversion from attack of both **SSIP(A)** and **SSIP(B)**.

The third possibility is for the SSIPs to fall apart with each ion moving its own way, leading to *free ions*. The only likely path for free ions is the attack of solvent on the carbocation, leading to the solvolysis product. This would take place with complete racemization. The return of free ions to SSIPs, CIP, or reactant is highly unlikely since the odds that these two ions will find each other are remote within a solvolysis reaction.

The upshot is that the $S_N1$ reaction as discussed in introductory organic courses is a simplification of reality. In actuality, the cleavage of the C–LG bond does not lead simply to free ions, swimming independently in the solution. Rather, a set of intervening types of ion pairs participate, and each of these type can react in a variety of ways. This set of ion pairs helps us to understand the variety of experimental results that include a range of stereochemical outcomes of the product along with racemization of the starting material.

The $S_N2$ mechanism is also a simplification, an idealization, of reality. While backside attack appears to be a major constraint, the degree of synchronicity of the bond making and bond breaking can vary widely. And certainly we observe reactions where multiple nucleophilic substitution pathways are occurring together.

Similar complexities appear upon careful examination of elimination reactions. The ion pairs I have discussed also play a role in the $E_1$ pathway. I want to discuss only one new aspect here: introducing a powerful experimental tool whose application to elimination reactions develops the notion of a spectrum of reaction mechanisms.

I have mentioned the use of isotopes a couple of times already. A key feature of isotopic substitution is that using a different isotope does not affect the underlying chemistry. Remember that chemistry is all about making and breaking chemical bonds, and that entails moving electrons. Isotopes have the same number of protons, which means that they have the same number of electrons. Furthermore, since neutrons carry no electrical charge, changing the number of neutrons in the nucleus will have no effect on the electrons. That's why we talk about isotopes having identical chemical behavior.

And this is mostly correct! Keep in mind, however, that isotopes have different masses, and so anytime that the motion of the atom itself is important, we might have to be concerned by the mass difference. Consider a few examples from the classical world. What's easier to do—pick up a box of cast iron pans or the same-size box filled with feathers? Obviously, a heavier object requires more work to pick up and move. Think about a drag race between a light car and a heavy truck. The extra weight of that truck will likely mean a slower start and longer time needed to get up to top speed.

Lastly, think of two identical springs hanging from the ceiling. Attach a bowling ball to one spring and a golf ball to the other. Pull the two springs down an equivalent amount and let go. Which spring will vibrate faster? Your intuition is probably correct—the golf ball, being much lighter, will vibrate much faster than the spring with the much heavier bowling ball attached to it. Instead of the bowling ball, attach a tennis ball. The golf ball is slightly lighter (46 g) than the tennis ball (58 g), and so it will vibrate faster, but not much faster. As the mass difference gets closer, the springs will vibrate more similarly.

We measure vibration in terms of frequency, that is, how many times the spring goes up and back to its original position per second. The vibrational frequency of a spring depends on the square root of the mass. In order to get the largest difference in vibrational frequency, we need as big a mass difference as possible.

The atoms in every bond will vibrate all the time. Even at absolute zero, the coldest possible temperature, the atoms will vibrate, creating what we call *zero-point vibration*. This is one of those bizarre manifestations of quantum mechanics. As temperature goes down, classical mechanics says that everything will move less and less; think of this as freezing out motion.

Classical mechanics holds that at absolute zero, all motion stops. One of the most famous components of quantum mechanics is the Heisenberg Uncertainty Principle, which prohibits complete knowledge of an object's position and momentum; essentially we cannot know exactly where an object is located *and* how it is moving. If an object stops moving at absolute zero, then we would know exactly where the object is and its movement—there isn't any; it's stopped. The Uncertainty Principle has been tested many times without any failure—so far. We have observed systems at temperatures just a small fraction of a degree above absolute zero, and atoms continue to vibrate!

Every bond is vibrating and has energy associated with that vibration. If we compare two bonds that differ just by having different isotopes, they have differing vibrational energy and different energy content. The bond with the heavier isotope, vibrating more slowly, will have less energy than the bond with the lighter isotope. If the reaction involves breaking that bond, then the activation barrier for the two systems will be different (see Figure 12.12) because of that difference in their initial vibrational energy. The black curve represents the potential energy for the reaction. Since every bond is vibrating, the molecule will not sit at the very bottom of the well, but rather will have some energy associated with the vibrations. The horizontal dotted line represents the energy of the reactant with the heavier isotope. It sits a bit lower in energy than the molecule with the lighter isotope, drawn as the dashed line.

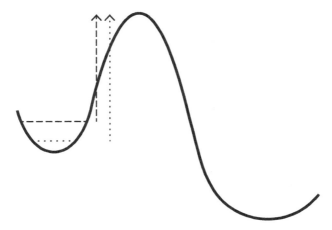

**Figure 12.12.** Potential energy diagram for bond breaking with an isotopic system. The dotted line denotes the heavy isotope and the dashed line denotes the light isotope.

The activation energy for the reaction is shown as the two arrows, the smaller dashed arrow indicating a lower activation energy for the lighter isotope system than for the heavier isotope system, shown by the longer dotted arrow. Since the activation energies are inversely related to the rates of reaction, this implies that molecules with the lighter isotope will react faster than molecules with the heavier isotope. We typically measure this as the *kinetic isotope effect* (KIE): the ratio of the rate of the lighter isotopic system to the rate of the heavier isotopic system: $KIE = k_L/k_H$.

Two major factors determine the magnitude of the kinetic isotope effect. The first factor is the difference in mass of the isotopes: the greater that difference, the greater the difference in their vibrational energies, leading to a larger difference in activation barriers of the light versus heavy isotope. Most isotopes differ by only a small number of neutrons. For example, each of the three isotopes of carbon has six protons and either six, seven, or eight neutrons. That means that the ratio of their masses is pretty small. That typically leads to small KIEs for reactions involving carbon isotopes, usually with KIE less than 1.1.

On the other hand, the three isotopes of hydrogen have large mass differences. The most abundant hydrogen isotope has just a single proton, with a mass of one ($^1H$). Deuterium has one proton and one neutron for a mass of two ($^2H$). Tritium, which is radioactive, has one proton and two neutrons, for a mass of three ($^3H$). With deuterium twice as heavy as normal hydrogen, we can anticipate some large isotope effects. In fact, $k_H/k_D$ can be upward of seven in some cases, with $k_H/k_T$ even larger still.

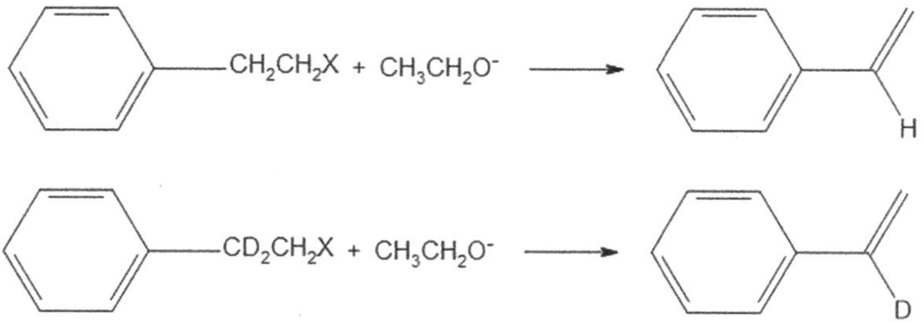

| X | $K_H/k_d$ |
|---|---|
| Br | 6.03 |
| N(CH₃)₃ | 2.98 |

**Figure 12.13.** Kinetic isotope effect in an elimination reaction.

The second factor contributing to the KIE is the degree of bond breaking that takes place in the transition state. The greater the extent of bond breaking in the transition state, the larger the kinetic isotope effect. This correlation will be useful in our further discussion of elimination reaction.

William Saunders examined the KIE of some elimination reactions, and the results of two examples are shown in Figure 12.13. When the leaving group is bromide, the KIE is very large, 6.03. (In fact, at colder temperatures, the KIE exceeds 7.) With a better leaving group, $N(CH_3)_3$, the KIE is half as large.

These compounds are both likely undergoing an $E_2$ reaction, but the different KIEs are telling us that the degree of C–H cleavage is different in the two reactions. The better leaving group $N(CH_3)_3$ needs less assistance, less "pushing," than the poorer leaving group Br⁻. The assistance in the bromine case comes from significant cleavage of the C–H bond, building up considerable negative charge on the carbon, which can then move to form the double bond and aid the bromine to exit. The differences in the transition states of the two reactions are shown in Figure 12.14. Note the difference in the lengths of the C–H bonds and C–LG bonds. In other words, there is a spectrum of $E_2$ transition states.

Other elimination reactions show almost no KIE. Rates studies and stereochemical studies indicate that there is spectrum of elimination mechanisms, not just within the $E_2$ realm. This can be depicted through the plot shown in Figure 12.15. Each axis depicts the distance of a breaking bond in the elimination reaction: the $x$-axis presents the distance of the C–LG bond, and the $y$-axis presents the C–H bond distance. The reactant is in the upper left corner and the product is in the bottom right corner, so in general the reaction proceeds from top left to bottom right.

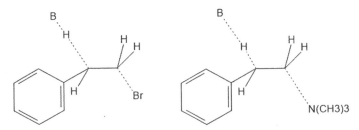

**Figure 12.14.** Transition states for $E_2$ reactions with bromine and $N(CH_3)_3$ as leaving group. Note the differing lengths of the C-H and C-LG bonds in the two transition states..

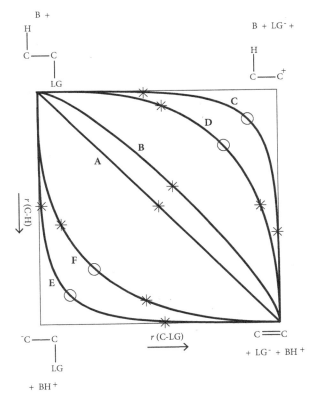

**Figure 12.15.** Diagram of the elimination reaction spectrum. Transition states denoted by * and intermediates by a circle. The $E_2$ mechanism is represented by curves A and B. The C and D are for $E_1$ reactions, and the E and F curves are for $E_{1cb}$ reactions..

The $E_2$ mechanism is perhaps the simplest to consider. If both bonds break synchronously, we have the A curve. The transition state (TS) is denoted by the asterisk. The TS is shown in the middle, but in fact the transition state could be earlier, with less stretching of both bonds, by indicating the asterisk moved toward the top left, or later, with more stretching of both bonds, by indicating the asterisk moved toward the bottom right.

The $E_2$ reaction is also possible where the two bonds break at different rates, meaning that in the transition state, one bond breaking has progressed farther than the other, as depicted in the TSs shown in Figure 12.14. In this case, the bonds break concertedly (in one step) but not synchronously. This is shown in curve B, where early on the C–LG group bond stretches more than does the C–H bond.

Note that in the $E_2$ mechanism there is one and only one TS and no intermediate. Also note that I have drawn in only two examples of $E_2$ pathways, but many other options are possible, lying both to the upper right and to the lower left of the A curve. This is what we mean by a spectrum of $E_2$ pathways.

The $E_1$ pathway is depicted as curve C in Figure 12.15. At the start, just the C–LG bond is stretching, reaching the first TS and then continuing to stretch, with little change in the C–H distance, to the intermediate, marked as a circle. The second step involves stretching of the C–H bond, through the second TS, and then to product. Again, other $E_1$ paths are possible, such as curve D. The positions of the TSs and intermediate have changed, depending on the leaving group's ability, the strength of the base, the acidity of the substrate, and other factors. One item to consider is the position of the intermediate along path C and D. Their main difference is in the C–LG distance. In both, the bond is broken. The intermediate on the C pathway has a very large C–LG distance, indicative of free ions. The C–LG distance is shorter in D than in C, suggestive of some type of ion pairing, either solvent separated or contact ion pairs.

A third distinct mechanism is also observed, depicted as curve E. This mechanism is often not included in introductory organic chemistry texts, yet many examples are known and well understood. In this mechanism, the first part of the reaction proceeds through stretching of the C–H bond, brought on by base attack, going through a transition state, and then to an intermediate carbanion. The C–LG bond is largely unaffected. In the second step, the C–LG bond is broken. This is explicitly detailed in Figure 12.16. This is the $E_{1cb}$ mechanism, as it involves the conjugate base as an intermediate. The $E_{1cb}$ mechanism comes in a few different varieties, and it too can follow different paths, such as curve F. Curves E and F differ mainly in the nature of the intermediate, again as free ions or ion pairs.

I hope that you recognize that both substitution reactions and elimination reactions can proceed via a spectrum of mechanisms. The world is not black or white. Substitution reactions do not neatly divide into the $S_N1$ or $S_N2$ mechanism. This richness of mechanism actually provides tools for synthetic chemists to play with reaction conditions, substituents, leaving groups, bases,

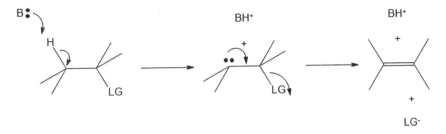

**Figure 12.16.** Reaction mechanism of the $E_{1cb}$ pathway.

and so on, to control the reaction outcome. What looks like an unworldly complication is actually a feature!

I also hope that this chapter-long digression helps you recognize why textbook authors provide simplified models as entry points into the field. These models, though not perfect, capture enough of the reality of the reactions to help students develop a chemical intuition that can later be fleshed out, gaps filled in, and subtleties explored as the student matures.

| X | $K_H/k_d$ |
|---|---|
| Br | 6.03 |
| N(CH$_3$)$_3$ | 2.98 |

# 13
# More Hard Truths

## Alkyls as Electron-Donating Groups

In this chapter, I present another example of chemical reality that is more subtle than what is presented in the introductory organic course or textbook. It's another case of where I thought I understood organic chemistry and where I was overconfident entering graduate school, but in fact I needed to learn how to be more inquisitive of assumptions.

This example involves a bedrock concept in all of chemistry: acidity. I have already introduced the concept of an acid; Bronsted and Lowry define an acid as a molecule that releases $H^+$. It is useful to note that while a hydrogen atom contains a proton in its nucleus and one electron moving about it, the hydrogen cation, $H^+$, is just a bare proton. That allows us to identify acids as *proton donors*, molecules that release a proton, or $H^+$.

The strength of an acid reflects just how many protons are released. We can define a generic acid as HA, where H is the proton to be released and A is the rest of the molecule. Hydrochloric acid might be the first example of an acid one sees in an introductory class; it is HCl, which breaks up into a proton $H^+$ and chloride $Cl^-$ in solution. Nitric acid, another powerful acid seen in introductory courses, is $HNO_3$; it breaks into $H^+$ and $NO_3^-$; so in this case A stands for $NO_3^-$.

The chemical reaction that defines an acid is $HA \rightarrow H^+ + A^-$. For the strong acids, the ones you hear about in high school chemistry such as hydrochloric acid or nitric acid, the reaction happens completely. If you were to toss a million molecules of HCl into water, essentially all million of these molecules would break apart, and you'd have a million protons, a million $H^+$, swimming about in that solution.

Not all acids are like that, though. In fact, most are not. Most acids will not completely break apart in water or in some other solvent. For example, acetic acid, the molecule that gives vinegar its particular smell and taste, is a much weaker acid. Only about 0.4% of the acetic acid molecules will dissociate and break apart into ions: a proton and an acetate anion. To indicate that a reaction does not go to completion, we use the double arrow: $HA \rightleftharpoons H^+ + A^-$. This

*Thinking Like a Physical Organic Chemist*. Steven M. Bachrach, Oxford University Press. © Oxford University Press 2023. DOI: 10.1093/oso/9780197640371.003.0013

double arrow indicates that the reaction is going forward and backward at the same time, but the rate of reaction forward is the same as the rate backward. That means we have a steady state with all three species present to some extent. We call this situation *equilibrium.*

Since the equilibrium is static, we can quantify the situation using the equilibrium constant, which is given as the concentration of the products divided by the concentration of the reactants. Recall that we denote concentration by putting brackets about the designation of the molecule. So, the equilibrium constant $K_a$ for the reaction of an acid is

$$K_a = \frac{[H^+][A^-]}{[HA]}$$

To determine the value of the equilibrium constant for an acid, we simply need to measure the initial concentration of the acid, and the proton concentration $[H^+]$ at equilibrium. In the past, this would have been measured using pH paper, a strip of paper coated with compounds that change color depending on the proton concentration. Nowadays, we have electrical probes that attach to a digital meter that directly reads out the pH, which is readily converted to concentration using the formula

$$[H^+] = 10^{-pH}$$

The carboxylic acids are organic compounds that are proton donors. A wide variety of naturally occurring materials are carboxylic acids, including the amino acids that make up proteins, aspirin (acetylsalicylic acid), and vitamin C (ascorbic acid. For our purposes here, I will focus on benzoic acid **1**, and its action as an acid (Figure 13.1).

Louis Hammett, a major figure in the creation of physical organic chemistry, developed a tool for identifying substituent effects. Substituents are

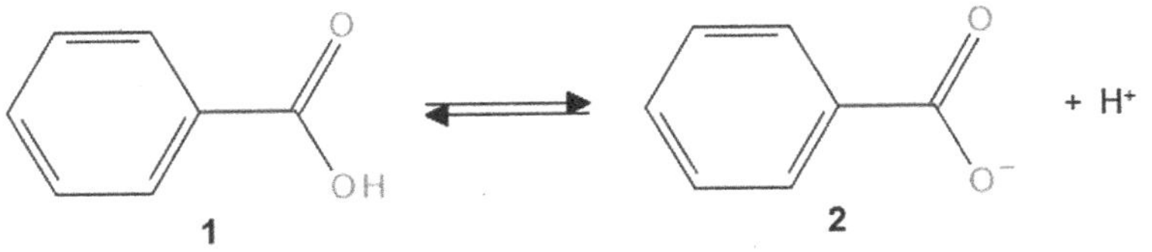

**Figure 13.1.** Acid reaction of benzoic acid **1**.

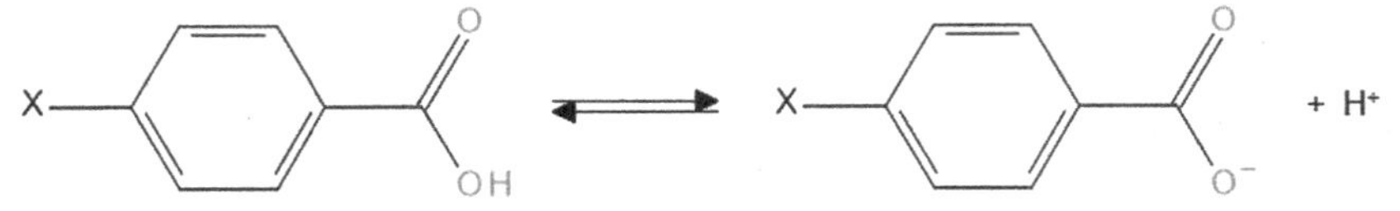

**Figure 13.2.** Acid reaction of substituted benzoic acid **3**.

atoms or groups of atoms attached to some larger fragment of the molecule. In the case at hand, Hammett explored a variety of different substituents attached to the phenyl ring opposite the position of the acid group, as in **3** (Figure 13.2). He examined the effect each substituent had on the acidity, namely, did the substituent make the molecule more or less acidic than the parent molecule **1**. (He actually quantified this effect, but the trend is all we need here.)

Substituents like bromide (Br) or nitro ($NO_2$) increase the acidity, while substituents like amino ($NH_2$) decrease the acidity relative to benzoic acid itself. These effects can be understood in terms of what the substituents can do (or not do) to stabilize the resulting anion. If the substituted anion is stabilized relative to the parent benzoate anion (**2**), then the acid reaction will take place to greater extent; that is, it's a stronger acid. If the substituted anion is destabilized relative to **2**, then the acid reaction will take place to a lesser extent; that is, it's a weaker acid.

How can a substituent stabilize an anion like **4**? Concentrating charge to a small location is destabilizing—you're forcing repulsive objects to be near each other. Spreading charge out over a larger volume, a process that chemists call *delocalization*, reduces that repulsion and stabilizes the anion. Recall that bromine and the other halogens, sitting on the very far right side of the periodic table, are eager to gain an electron. With bromine as a substituent on **4**, it will polarize the electron distribution on the neighboring phenyl ring (that's what we call a benzene ring when it has substituents attached to it), moving some electron density toward itself. This leaves the opposite side of the ring with lessened electron density or a slight positive charge. That partial positive charge is adjacent to the carboxylate group that has the negative charge, allowing for some of the excess negative charge to bleed over (delocalize) into the phenyl ring.

A similar sort of delocalization happens with substituents like the nitro group. To explain how the nitro group stabilizes the anion, let's examine the two resonance structures of the anion **5** (Figure 13.3). The left resonance structure has an overall neutral nitro group and neutral phenyl ring. The right resonance structure moves electron density out of the phenyl ring onto the

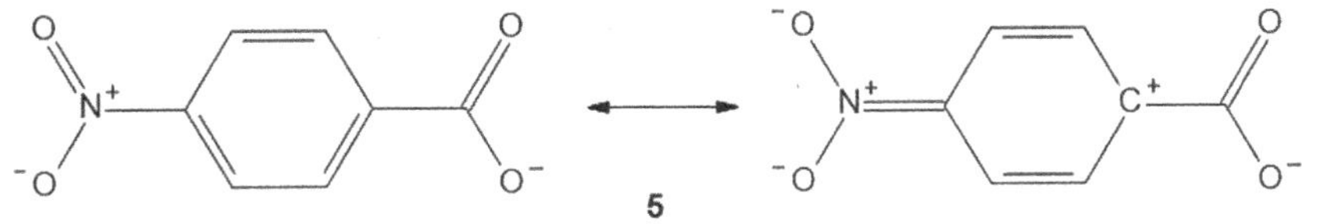

Figure 13.3. Resonance structures of anion 5.

Figure 13.4. Resonance structures of anion 6.

electronegative oxygen atoms of the nitro group. That movement, like that described above for bromine, polarizes the phenyl group, placing some positive charge adjacent to the carboxylate anion. The carboxylate can then shift some of its excess negative charge onto the phenyl ring. This leads to delocalization of the negative charge over a larger volume than in 2, for a net energetic stabilization.

Both bromine and the nitro group act to move electron density from the ring to themselves. We call these types of substituents *electron-withdrawing groups*. These groups will increase the acidity when attached to benzoic acid and stabilize any substrate that has some negative charge.

How about a rationale for the amino group that reduces the acidity when attached to benzoic acid? Again, let's make use of resonance structures, this time for 6 (Figure 13.4). The left resonance structure has a neutral amino group and phenyl ring. However, the right resonance structure features a positively charged amino group, and the phenyl ring carries a negative charge, placed on the carbon that has the carboxylate. Note that this resonance structure places two negative charges very close to each other, which clearly presents an unfavorable situation leading to a less stable anion. Since the anion is less stable, the amino-substituted benzoic acid will be less acidic and less likely to lose the proton to form this less stable anion 6.

The amino group has acted to give some electron density to the neighboring ring. Amino and related groups are therefore called *electron-donating groups*. These groups will reduce the acidity when they are attached to benzoic acid. They will also destabilize any system that carries some negative charge.

What effect might a carbon chain substituent, like a methyl group $CH_3$ or ethyl group $CH_3CH_2$ have on the acidity of benzoic acid? Will they be electron-donating groups and reduce the acidity or electron-withdrawing groups and enhance the acidity? Hammett's studies indicate that carbon substituents reduce the acidity of benzoic acid, though not to the extent caused by an amino group. They are weak electron-donating groups.

This result squares with what we have previously seen regarding the stabilities of cations. In the discussion of the $S_N1$ mechanism in Chapter 5, I presented evidence for the trend in stability of carbocations, as shown in Figure 5.5. In essence, the more substituted the carbocation, the more stable it is. Referring back to Figure 5.5, if we consider the central carbon that carries the positive carbon as the core component, then we can consider each methyl group as a substituent. Moving from primary to tertiary carbocation means adding ever more methyl groups. Each of these methyl groups is an electron-donating group. Each group will move some electron density onto that central cation, helping to stabilize the overall ion by effectively delocalizing the positive charge onto each added methyl group. So, while an electron-donating group will destabilize an anion, it will stabilize a cation.

That's pretty much the standard presentation of the concept of the carbon substituents (*alkyl groups*) as electron donating in every introductory organic chemistry textbook and in probably most introductory organic chemistry classes. This rationale is used to understand many trends, notably the following example.

Alcohols are compounds with a hydroxyl group OH, and can generically be represented as ROH. That hydroxyl hydrogen can be released, making alcohols somewhat acidic, that is, through the reaction $ROH \rightleftharpoons H^+ + RO^-$. The acidity of many small alcohols has been extensively studied. In water solution, the relative acidity of the smallest alcohols are shown in Figure 13.5. The smallest alcohol, methanol, is the most acidic. Each methyl group added to the carbon that carries the hydroxyl group decreases the acidity.

This trend can be understood by again looking at the stability of the resulting alkoxide anion $RO^-$. If every methyl group is electron donating, then each added methyl pushes more and more electron density toward the carbon adjacent to the oxygen atom with the negative charge. That is a destabilizing interaction, bringing negative charges near each other. So, the smallest alcohol,

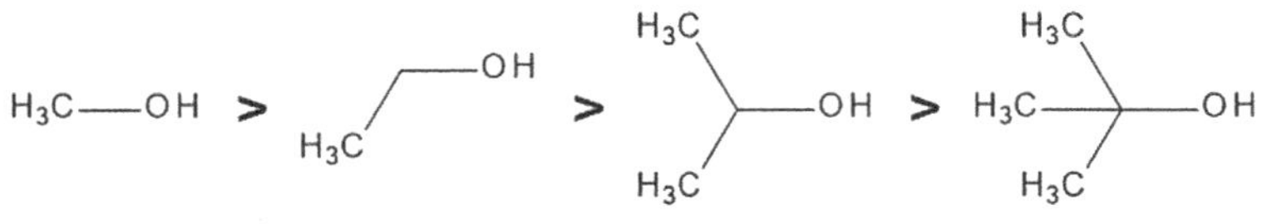

**Figure 13.5.** Trend in solution phase (water) acidity of the small alcohols.

methanol, with the fewest electron-donating groups will be the strongest acid. The largest alcohol, *t*-butanol, will be the least acidic because its anion is destabilized by those three methyl groups.

I can still recall the lecture during my first semester as a graduate student in which our professor presented the work of John Brauman, a physical organic chemist working across the San Francisco Bay at Stanford University. Brauman measured the acidity of these same alcohols but in the gas phase. Molecules in the gas phase are isolated and independent. Observations of properties of molecules in the gas phase are considered to probe their true nature, unperturbed by any neighboring solvent molecules or other things swimming about the solution.

I remember this lecture because the acidity trend is completely reversed in the gas phase from the solution results! Methanol is the least acidic and *t*-butanol is the most acidic in the gas phase (Figure 13.6).

OK, so how do we make sense of that trend? Delocalization of charge is still the key to the story. In the gas phase, when a molecule is all by itself, concentration of charge is difficult to bear. Rather than thinking of the methyl group as an electron donor, in the gas phase, what matters is just sheer size. The methoxide anion $CH_3O^-$ is very small, and *t*-butoxide $(CH_3)_3CO^-$ is so much larger. That excess charge can be distributed across a much larger volume in the bigger molecule, stabilizing the *t*-butoxide anion, and making it easier to form than methoxide.

If that's true, then what do we make of the electron-donating group argument? The answer to this question was what made that particular lecture back at Berkeley so memorable. The gas-phase experiment is telling us about the inherent nature of the alcohols: that bigger alcohols will be more acidic. The reverse of this trend in solution cannot be attributed to some electron-donating ability; electron-donating ability should be an attribute of the molecule itself and observable in the gas phase. But in the gas phase, where the molecule is all by itself, the electron donation concept predicts the entirely wrong trend. We must reject the concept that alkyl groups are electron donating!

Instead, the only possible option is that the trend in solution phase acidity of the alcohols is somehow attributable to the solvent. I can offer a couple

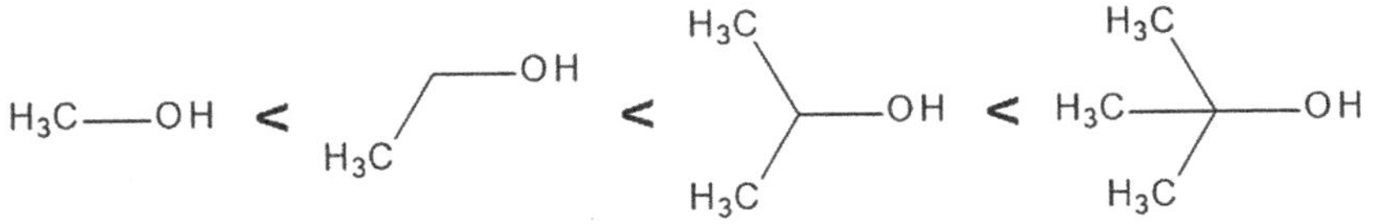

Figure 13.6. Trend in gas-phase acidity of the small alcohols.

of roles that solvent may play. First, hydrogen bonding between the solvent (water) and the oxyanion is certainly playing a significant role in stabilizing the anion. The alkyl portion of the molecule is *hydrophobic*, not capable of effectively interacting with water. (That's why oil and water do not mix but instead separate into layers.) As the alkyl portion gets larger, more water is ever more excluded from approaching the oxide end of the anion, diminishing the hydrogen bonding. Second, ever-larger hydrophobic alkyl groups prohibit the water from getting close enough for favorable dipolar interactions.

The explanation for carbocation stability must also be called into question. We reject the notion of alkyl groups for acidity, and we should also be hesitant to attribute the carbocation stability trend to this notion too. The cation stabilities are also likely attributable to size and solvation ideas.

These results forced me, as a beginning graduate student, to recognize that simple notions may not suffice. Assumptions need to be questioned. Is a solvent really a benign medium, needed just to get reactants into solution so that they can move freely about? Do trends observable for one reaction transfer to another reaction?

Are introductory organic chemistry textbooks to be trusted at all? This is a difficult question to answer. Organic chemistry is a rich, subtle field. It is beautiful and complicated and difficult to grasp, especially to newcomers. Learning organic chemistry is like learning a new language, with unfamiliar words and a different grammar—though in this case the grammar is a set of rules and logic structure different from anything a student has probably seen before.

Textbook authors need to present the material in some manageable form: providing some glimpse of the subtleties but without overwhelming the students. We simplify substitution reactions to two extreme cases ($S_N1$ and $S_N2$). This provides some introduction to nuance, the interplay between degree of substitution, nucleophile strength, leaving group ability, and solvent. But we omit the complexities of ion pairs and a continuum of mechanism. We continue to teach that alkyl groups are electron donating because it gets us to the correct prediction, and we neglect to teach that the idea is built upon quicksand.

I don't believe this approach is intellectually dishonest. In fact, I believe it reflects how most education proceeds. The American Civil War was not only about slavery, but that cause helps animate interest in the history of the era. As students get older, they are prepared to learn about the other causes, notably, states' rights, economics, and expansionism, and then to dive back again into slavery to read about its incredible cruelty and widespread observance, even in the North.

Simplified models provide an entrance into new material. Continued study fleshes out the subtleties, fills in the gaps, and refines the models. A mind needs to be prepared to learn and then needs to be guided through the complex terrain. Care must be taken that the student not drown in all of the details or lose the forest for the leaves. It's treacherous going.

The lesson is that we should never stop inquiring, never accept the first treatment as the final version. Assumptions need to be subjected to scrutiny, and we must be ready to cast off old notions when new data emerge.

# 14

# Addition Reactions

## Kinetic and Thermodynamic Control

Synthetic organic chemists want options, such as methods that will transform molecules in one direction *and* methods that will reverse that transformation. In earlier chapters, I presented the elimination reaction, transforming a molecule having a hydrogen atom and some leaving group on an adjacent position into an alkene, a molecule with a carbon–carbon double bond. Can we produce the opposite reaction: transform an alkene by adding a hydrogen and some other group (sometimes even two groups, neither of which is a hydrogen atom) across the double bond?

The answer is "yes," through what we call an *addition reaction*! Generically, an addition reaction adds the molecule A–B across a double bond, as in Figure 14.1. More specific are the two examples shown in Figure 14.2.

Of particular note is that the two examples of addition reactions in Figure 14.2 yield only a single product, which is true of the vast majority of addition reactions. One might have expected the addition to go in two different ways. Looking at the generic reaction in Figure 14.1, we might expect the product shown, along with the product where the A and B groups are swapped, such that the A group could end up on either the left or right carbon of the double bond. Instead, only one product is obtained. When we have the possibility of molecules combining in more than one orientation but only one of these possibilities is actually produced, we refer to this reaction as a *regiospecific reaction*.

This regiospecificity of addition reactions was noted in the late nineteenth century by the Russian chemist Vladimir Markovnikov. As with many Russian names, transliteration of his name often results in different spellings, with Markownikoff being a common alternative. Markovnikov noted this regiospecificity and supplied a means for predicting the orientation, which is now known as *Markovnikov's rule*. This rule can be stated in many ways, and I prefer the following: the more electronegative addend, that is, either A or B, will end up on the more substituted carbon of the original double bond. So, in the top example of Figure 14.2, chlorine is more electronegative than H,

*Thinking Like a Physical Organic Chemist.* Steven M. Bachrach, Oxford University Press. © Oxford University Press 2023.
DOI: 10.1093/oso/9780197640371.003.0014

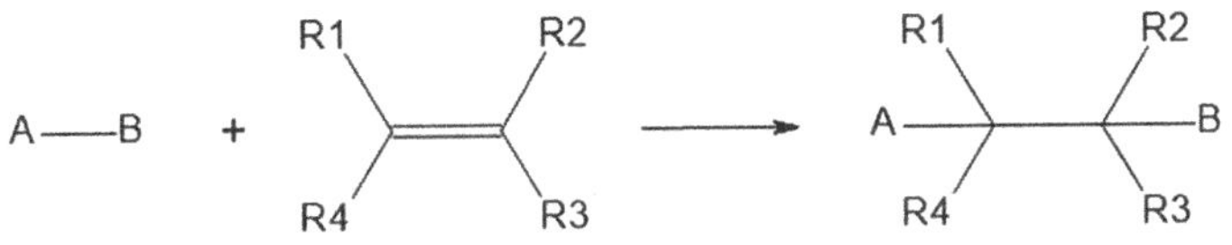

**Figure 14.1.** Generic representation of the addition reaction.

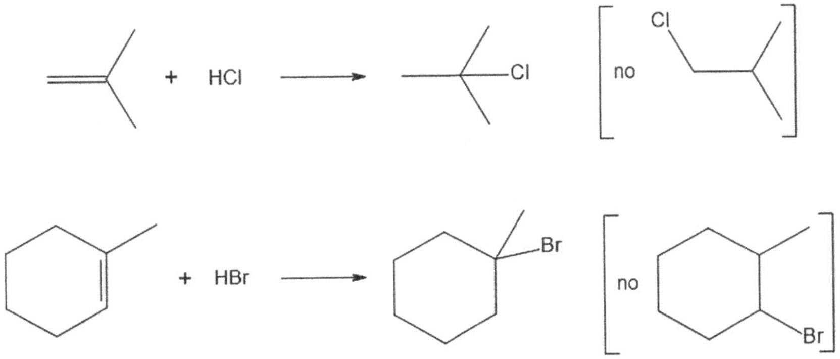

**Figure 14.2.** Two examples of addition reactions to alkenes demonstrating their regiospecificity.

and it ends up on the more substituted carbon, which is the right one since it has two substituents while the left carbon has no substituents. In the bottom example, bromine, more electronegative than hydrogen, ends up on the more substituted top carbon of the double bond.

What is the rationale for this regiospecificity? Markovnikov, of course, offered no such answer since mechanisms were not discussed for many more decades; even the electron had not yet been discovered at his time! Let's consider the bonds that are made and broken in the top reaction of Figure 14.2. The double bond is broken, as is the H–Cl bond, and the C–H and C–Cl bonds are made. We might next consider the possible order of making/breaking these bonds, but a more fruitful approach is to consider the nature of the H–Cl bond and the double bond.

HCl is a strong acid, and it is best considered as $H^+$ and $Cl^-$. The carbon–carbon double bond is electron rich, with four electrons shared in the region between the two atoms. The species that is likely to attack an electron-rich region is something that is positively charged, like the $H^+$ from HCl. We call a molecule that is seeking negative charge an *electrophile*. That suggests that the first step of the reaction mechanism is the formation of the C–H bond.

The bond can be formed to either carbon, which are depicted in the in two intermediates **1** and **3** shown in Figure 14.3.

Since an $H^+$ has been added, both intermediates are carbocations. Their most important difference is that **1** is a tertiary carbocation, while **3** is a primary carbocation. The next step is for chloride to attack the cation center. Reaction of carbocation **1** leads to product **2**, with the chlorine on the more substituted carbon of the former double bond. Attack of carbocation **3** gives **4**, having the chlorine on the less substituted carbon.

Experiment tells us that the only observed product is **2**. Can the mechanism of Figure 14.3 explain that result? The two possible reaction pathways differ in the type of intermediate carbocation that is formed. We've seen these types of carbocations before, way back in Chapter 5 where I presented the $S_N1$ reaction. Carbocation stability increases with substitution, telling us that intermediate **1** will be much more stable than intermediate **3**. The reaction proceeds only through the much more stable tertiary cation, leading to product **2** only.

The second addition reaction presented in Figure 14.2 can be explained in the same way. The two possible intermediates are a secondary and tertiary carbocation. Again, the tertiary carbocation is so much more stable than the secondary cation that the only product is the one where the bromine attacks the tertiary cation. This is the observed product, where bromine is connected to the more substituted carbon of the former double bond.

This discussion is actually just a setup for the next reaction, which presents an important concept in physical organic chemistry. In Chapter 10, I mentioned the notion of a thermodynamic product and a kinetic product in the discussion of elimination reactions. Here, I will develop this concept further as another tool enabling organic chemists to control reactions.

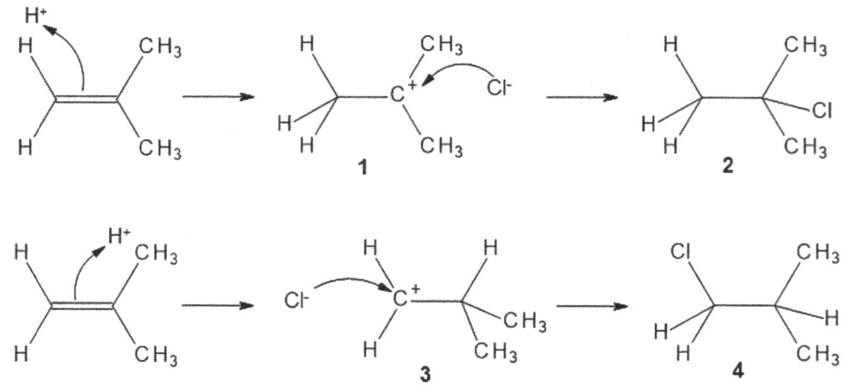

**Figure 14.3.** Mechanism for addition reactions.

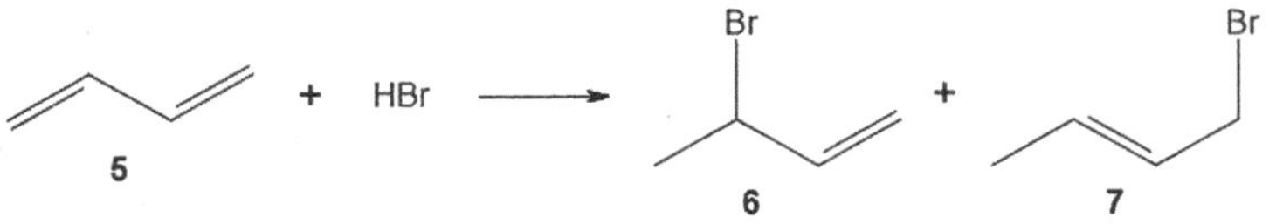

| | –78 °C | 90 | 10 |
| | 0 °C | 71 | 29 |
| | 20 °C | 44 | 56 |

**Figure 14.4.** Addition of HBr to 1,3-butadiene

The reactant **5** shown in Figure 14.4 is an example of a diene, a molecule with two double bonds. It is also a *conjugated* diene, where the double bonds are separated by one single bond. Conjugated dienes can undergo addition reactions just as regular alkenes do, to give the product **6**. The hydrogen adds to the terminal carbon, setting up the cation on the second carbon, and then a bromide anion attaches to that position.

However, **6** is not the only product observed; **7** is also identified as a product of this reaction. **7** comes about through addition of the hydrogen to one of the terminal carbon atoms and addition of bromine to the other terminal carbon. Compound **6** is referred to as the *1,2-product*, with the addends (H and Br) attaching to adjacent carbons, while compound **7** is the *1,4-product*, where hydrogen adds to one terminal carbon and bromine attaches to the fourth carbon of the diene.

How can we explain that both 1,2- and 1,4-addition take place in the reaction of conjugated dienes? The key element is the structure of the intermediate carbocation. The proton acts as the electrophile and attaches to the terminal carbon, producing the secondary carbocation **8a**, as shown in the first step of the mechanism given in Figure 14.5. The bromide anion can then attack the second carbon, the one carrying the positive charge—a sensible thing for an anion to do—creating product **6**.

The description of the carbocation, however, is incomplete: a second resonance structure, **8b**, is necessary to properly describe this cation. That second resonance structure implies that the positive charge is spread between the second and fourth carbon. The spreading of charge is always stabilizing, and so this intervening carbocation **8** is particularly stable and readily formed. It means that the bromide ion can also attack the fourth carbon because it, too, carries positive charge. That leads to the second product, the 1,4-product, **7**. So, the mechanism presented in Figure 14.5 accounts for the two products.

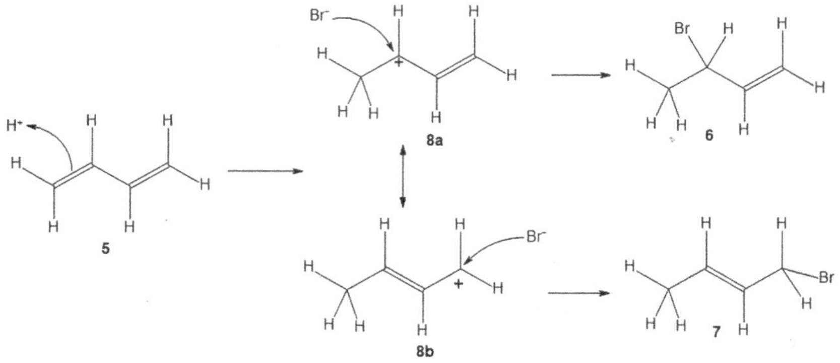

**Figure 14.5.** Mechanism for the addition of HBr to 1,3-butadiene.

There are unanswered questions, however. At low temperatures, the 1,2-product is made to a much greater extent than is the 1,4-product. As the temperature rises, the distribution shifts to prefer the 1,4-product (see Figure 14.4). If we consider both of the two resonance structures for **8** to be necessary to describe the cation, why aren't the two products formed in equal, or nearly equal, amounts? Why is there a temperature dependence in the product distribution?

To answer these questions, we must first recall what the major role of heat (temperature) is in a reaction. Every reaction has some energy barrier that must be crossed. As I discussed in many previous chapters, this barrier arises for both enthalpic and entropic reasons. Enthalpic barriers arise from the energy needed to partially break and make the bonds that are changing in the reaction. The entropic barrier arises from some organization that is necessary within a reaction, such as bringing two reactant molecules near each other. The energy required to surmount the reaction activation barrier is most often provided through heat. Increasing the temperature means that the molecules are moving more rapidly, usually through some combination of translation (moving left or right, say), rotation (spinning about), or vibration (bonds stretching and contracting). This added kinetic energy can then be transferred into climbing the barrier and facilitating the reaction.

When a reaction barrier is the key element of the story, we need to consider its kinetics, the rates of reaction. A small barrier means a fast reaction, whereas a large barrier means a slow reaction. Increasing the temperature, adding energy to the system, allows the reactants to climb over larger barriers. So, in examples like this one, the addition to a diene, where one product is favored at low temperature and the other product is favored at high temperature, we have a prime case pointing to different reaction barrier heights.

To understand the reaction shown in Figure 14.4, we will construct a reaction coordinate diagram. The first step of the reaction, as shown in Figure 14.5, is the addition of $H^+$ to 5 to form the intermediate cation 8. This step has a barrier, passing through the transition state we'll designate $8^\ddagger$. This step is endothermic, with a gain in energy. This step is represented in Figure 14.6 as the solid black line.

At low temperatures, the 1,2-product 6 is favored over the 1,4-product 7. That means that the barrier to form 6 is lower than the barrier to form 7. Let's modify our reaction coordinate diagram to include the curve (the dotted line) for the formation of 6, passing through transition state $6^\ddagger$. For now, let's just draw in a dashed line to indicate the energy of the transition state to 7, called $7^\ddagger$. Figure 14.7 captures all of this information.

Does it make sense that the transition state for 1,2-addition is lower than the transition state for 1,4-addition? Here's a possible answer. When the proton comes in to attack the terminal carbon of 5, the bromide is probably

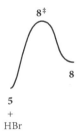

**Figure 14.6.** Potential energy surface for the first step in the addition of HBr to 1,3-butadiene 5.

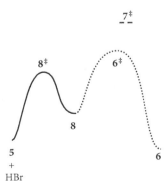

**Figure 14.7.** Further defining the potential energy surface for the addition of HBr to 1,3-butadiene.

nearby, figuring on balancing the charge. That would position the bromide close by to attack the second carbon but far away from the fourth carbon. One might imagine some reasonable enthalpic barrier to move the bromide away from the attacking proton, along with some entropic barrier to arrange the bromide nearer to the fourth carbon.

The experimental data indicates that increasing temperature shifts the product distribution in favor of the 1,4-product. Increasing the temperature means that we are providing more energy to the molecules and so *all* reaction rates should increase. As shown in Figure 14.7, that suggests that the rate to produce 7 should increase, which means more of it should be produced. That jibes with the experiment. However, the rate to produce 6 should also increase and remain faster than the rate to produce 7—the barrier to 6 is less than the barrier to 7. So if all we have is what's in Figure 14.7, we cannot explain the shift in production to 7 over 6.

The missing ingredient is that reactions that can take place in the forward direction can also take place in the reverse direction. This is the *principle of microscopic reversibility*: reactions proceed in both directions—forward and backward—through the same transition state. At the start of a reaction, when only reactants are present, the reaction takes place only in the forward direction, making products. But as products accumulate, the backward reaction begins to occur. Over time, the forward reaction slows down, as there is less reactant present, and the backward reaction speeds up, as more and more product are present. At some point, equilibrium is reached: the forward and backward reaction rates are identical. Reaction still occurs, but the concentration of reactant and product no longer changes; for every forward reaction that occurs, a complementary backward reaction also takes place.

To motivate the role of the backward reaction in solving the puzzle of diene addition, I will digress with an analogy. A young couple are contemplating purchasing their first home. Their desires far outweigh their bank balance: a large yard for the dogs to run around in; a swimming pool; four bedrooms; a media room—most of these amenities will have to wait. So they decide on a long-term plan. They will buy a starter home now, live in this home for a few years, build up some equity, and then sell this home to buy a larger home. They will repeat this process one or two more times, and then eventually they will end up in their dream home. This repeated buying and selling is akin to a reaction going forward and then backward, repeated a few times, but eventually settling in the desired end product: the perfect dream home!

The last consideration we need to make to our reaction coordinate diagram can be drawn from our home-buying analogy. The dream home is the best place to end up. It is analogous to the lowest energy structure—the best energetic possibility. That implies that product 7 must be lower in energy than 6. Does that make sense? As I mentioned in the earlier discussion of elimination reactions, alkene stability increases with substitution about the double bond, so let's examine the degree of substitution in each product. In 6, there is only one substituent, and it's attached to the second carbon; the terminal carbon of the double bond has only hydrogen atoms attached to it. Now the double bond in 7 has two substituents, one attached to each carbon of the double bond; therefore, it is more stable than 6. We can finalize our reaction coordinate diagram by adding 7 at an energy below 6 and drawing in the reaction path from 8 through 7‡ to 7 as the dashed line in Figure 14.8.

Using the complete reaction coordinate diagram of Figure 14.8, let's consider what happens at low temperature after formation of the intermediate carbocation. At low temperature, there is enough energy for a reasonable rate of crossing over 6‡, creating the 1,2-product 6. However, there is insufficient thermal energy for any significant rate to cross 7‡, and so little 7 is produced.

Now consider increasing the temperature. Remember that the overall effect of that added heat is for *all reaction rates to increase*. Both 6 and 7 are formed more readily. But now we also need to consider the backward reactions. With the added heat, the reversion of 6 to the carbocation 8 is accelerated. That back reaction means a decrease in 6 and more 8, which now has the opportunity to proceed to 6 or 7. This means that some of 6 formed early on will disappear and then go on to form 7. That's how the product distribution can shift from 6 to 7. Now to be fair, we should also consider the back reaction returning 7 to 8, but this reaction will be by far the slowest of all, since it has the largest activation energy. Regardless, of what temperature we are at, reversion of 7 will always be the slowest process. In the language of our analogy, the couple

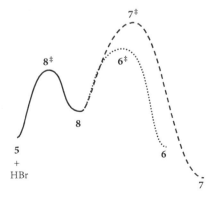

**Figure 14.8.**  Full potential energy surface of the addition of HBr to 1,3-butadiene.

are ready to buy and sell over and over again (essentially sampling **6** over and over) until they purchase their dream house (product **7**); they'll never leave that home!

Perhaps we can get a better handle on this through simulation of this reaction. Figure 14.9 presents the concentrations of carbocation intermediate **8** and the two products **6** and **7** as a function of time. At the start, we have just the carbocation. In the early few moments, **6** is formed much faster than is **7**. That's because of the much smaller activation barrier to make **6** than **7**. As time progresses, however, **6** begins to disappear; the back reaction takes **6** back to **8**. The back reaction for **7** has such a large barrier to surmount that little reversion takes place. Over time, the reaction keeps siphoning off **6** and heading it over the **7**, until an equilibrium is met, with a small amount of **8**, more **6**, and mostly **7**.

We can use Figure 14.9 to consider our diene addition reaction under low and high temperature. At low temperature, all the rates are slow, and the far right of this graph has a time that is centuries into the future. The longest one might typically run a reaction is overnight or maybe a day, and so the resulting situation is way over to the left of the graph—mostly **6** and a little bit of **7**. For high temperatures, the rates are faster, and the far right of the graph indicates a time of a few minutes. Under that circumstance, we observe mostly **7** and a smaller amount of **6**.

When we have a reaction with two or more possible products, the product formed through the lowest barrier is called the *kinetic product*. The kinetic product is the one formed fastest. In our example here, the barrier to the

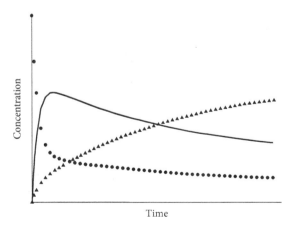

**Figure 14.9.** Plot of the concentrations of **6** (solid line), **7** (triangles), and **8** (circles) in the addition of HBr to 1,3-butadiene.

1,2-product **6** is smaller than the barrier to the 1,4-product **7**, so it is the kinetic product. When we have a reaction with two or more possible products, the most stable product, the one with the lowest energy, is called the *thermodynamic product*. Given enough time, the thermodynamic product will also win out—but that might require eons! In our example, **7** is more stable than **6**, so it is the thermodynamic product.

We can utilize temperature as a means to select the kinetic or thermodynamic product. At low temperature, only the lowest barriers will be crossed at any appreciable rate, and then we will select for the kinetic product. At high temperatures, we can ensure that all forward barriers will readily be crossed, and we will then obtain the most stable product, the thermodynamic product. This situation provides us with *kinetic control* (low temperature) or *thermodynamic control* (high temperature). This is a very important tool for synthetic chemists to wield, enabling the selection of the product they desire.

Unfortunately, not all reactions with multiple products exhibit the conditions for this selectivity. We need the situation shown in Figure 14.8: a low barrier leading to one product, but the other product is more stable. Often, we are faced with a reaction that has the energy diagram shown in Figure 14.10. In this case, the transition state leading to P1 is lower than the transition state leading to P2, making P1 the kinetic product. Additionally, P1 itself is more stable than P2, so P1 is also the thermodynamic product. In cases like this, where one product is both the kinetic and thermodynamic product,

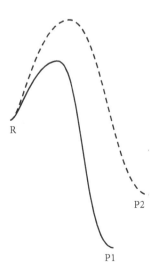

**Figure 14.10.** Potential energy surface where the kinetic ad thermodynamic products are identical.

temperature variation will not allow us to appreciably make the other product P2. P1 will be favored at all temperatures. If P2 is what we desire to make, we will need to find an alternative reaction or a catalyst that selectively lowers the barrier for making P2.

Kinetic and thermodynamic control have been useful tools for more than fifty years. Textbooks are replete with discussions and examples of this type of selection. Over the past decade or so, and completely out of the blue, emerged a third control mechanism. It is not as broadly applicable, but it surely is interesting. And it requires us to delve into the weird world of quantum mechanics, so let's tackle that in the next chapter!

# 15
# Quantum Mechanical Tunneling

Science education can focus almost entirely on discoveries situated in the past. In physics, we are taught about Kepler, Newton, Faraday, Maxwell, Einstein, Heisenberg, and Bohr. In chemistry, we learn about Dalton, Lavoisier, Fischer, Curie, Diels, Seaborg, Woodward, and Brown. In biology, we are introduced to Linneaus, Lemarck, Mendel, Darwin, Chargaff, Watson and Crick, Monod, and Peretz. It is as if science is carved in stone, as ancient as the Greeks, as fossilized as the dinosaurs.

Nothing can be further from the truth, of course. Science remains a fruitful, active endeavor that continues to enrich our lives. Consider what science has accomplished in this time of the Covid-19 pandemic. In about 14 months, scientists identified the virus, sequenced its genes, devised more than half a dozen vaccines, brought these vaccines into production, and inoculated billions of people across the globe.

This chapter and Chapter 18 describe ongoing experiments and interpretations that demonstrate some of what is exciting physical organic chemists today. Science is a continuing practice, and its modern state is just as evolutionary, if not revolutionary, as the highlights associated with many of the scientists mentioned above. The story in this chapter applies one of the strangest notions of quantum mechanics to chemical reactions, leading to the development of a third way that chemists can control reactions.

Quantum mechanics is perhaps the component of physics that has most captured the imagination of scientists and laypeople alike. It is just so strange, so unlike our normal experience, that it engenders wonder and speculation that appeal to science fiction writers, novelists, and scientists. How many episodes of *Star Trek* or *Star Wars* use some aspect of quantum mechanics as a jumping-off point?

Among the weird attributes of quantum mechanics, the concept of quantum tunneling may be the strangest of them all. In the classical world, a bedrock core principle is that barriers can only be crossed if the person or object has sufficient energy to climb over it. Imagine falling into a well with shear sides, like the pit in Buffalo Bill's basement in *The Silence of the Lambs*. Your only way out without any external assistance is for you to jump high enough

*Thinking Like a Physical Organic Chemist*. Steven M. Bachrach, Oxford University Press. © Oxford University Press 2023.
DOI: 10.1093/oso/9780197640371.003.0015

to grab onto the lip of the well. You need to supply the energy to overcome the gravitational pull that keeps you at the bottom of the well. If the lip of the well is 20 ft above your outstretched hand, and you are very athletic, you might jump up about 4 ft, but you're still woefully short of the top. Your hand remains many feet away from the top. You are trapped. The only way you are escaping that well is with some outside help. Someone lowers down a rope and you climb up. The rope is in effect a catalyst, providing an alternative pathway for you to convert the potential energy in your muscles into kinetic energy to climb up. Or someone drops a bucket down the well and pulls you up, using their potential energy to provide the kinetic energy needed to lift you out of the well.

Large barriers have thwarted human travel and progress since the dawn of time. Traversing mountains is tough business, and humans undoubtedly learned quickly to do so by crossing over the pass, the lowest point along the range separating one from the destination on the other side. As technology progressed, society came up with another solution—dig a hole through the mountain! The resulting tunnel avoids having to go up and over the pass, making the trek considerably easier and, presumably, faster. The first time I visited Colorado as a child in the late 1960s, we crossed the front range of the Rocky Mountains by going over Loveland Pass. This is a beautiful mountain drive, twisting and turning up the mountainside, eventually cresting at 11,990 ft (3,655 m) above sea level. Some ten years later, on my travel to graduate school in California, I crossed the mountains by driving through the Eisenhower Tunnel, a nearly 1.7-mile-long tunnel on Interstate 70 at 11,158 ft (3,401 m) above sea level. It's a shorter route, being straight through the mountain and saving about 800 ft of climb, which my underpowered Ford Pinto certainly appreciated!

Crossing the Alps is so important to the economy and society of the Swiss people that they have built three tunnels connecting the cantons of Uri in the north and Ticino to the south. The first tunnel, built in 1882, was the Gotthard Tunnel. This nearly 10-mile-long rail tunnel avoids having to climb over the Gotthard Pass at 6,909 ft (2,106 m)—see Figure 15.1a. The second tunnel, the Gotthard Road Tunnel, was constructed in 1980 for cars (Figure 15.1b). The third tunnel, the Gotthard Base Tunnel, is truly an astonishing engineering feat. Completed after seventeen years of work in 2016, this 35.5-mile-long rail tunnel is the longest and deepest tunnel in the world (Figure 15.1c). Its construction cost about $12 billion and required removal of over 28 million tons of rock. Each of these tunnels provides a shorter path and requires less climbing than is necessary with the previous alternatives.

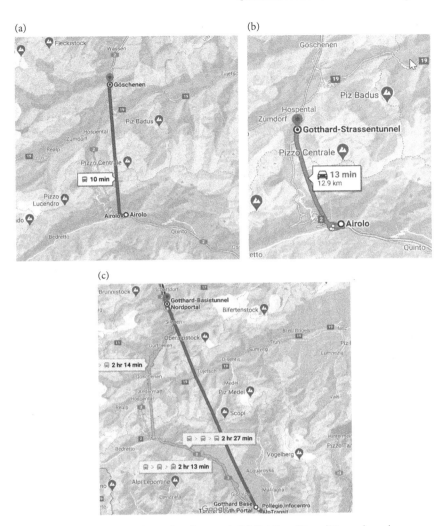

**Figure 15.1.** Maps of the (a) Gotthard Tunnel, (b) Gotthard Road Tunnel, and (c) Gotthard Base Tunnel.

I'm sure you see my point. You are blocked by some great impediment, be it the sheer walls of a well, the Rocky Mountains, or the Alps. Getting across will require a significant amount of energy or some external assistance. Through tremendous expenditure of energy, time, and money, a tunnel can be formed that will allow the next person to cross with great alacrity.

Imagine a surface that looks like the one presented in Figure 15.2. There is a well to the left with a very high sheer wall on the far left edge and a barrier to the right. Past the barrier is a lower-energy region. The best place, the lowest

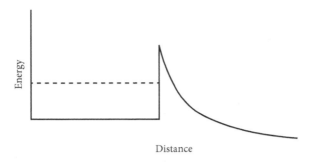

**Figure 15.2.** Model potential energy surface for alpha decay. The dashed line denotes a particle with an energy that would be trapped, according to classical mechanics, in the well.

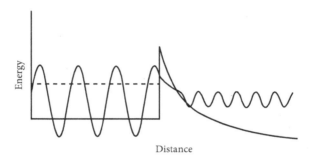

**Figure 15.3.** Wavefunction (in blue) for an alpha particle superimposed on its potential energy surface.

energy location, for an object on this surface is over to the right side. Now imagine that the object has the energy shown as the dashed line and the object is placed in the well to the left. In the classical world, that object is forever stuck in the left well. In the absence of any energy added to the object, there is no way for the object to surmount the barrier and get to the lower-energy region on the right.

Now what does quantum mechanics say about a particle residing in a potential energy surface like this one? In the quantum world, particles behave like waves, and the wave can penetrate into the barrier. Some of that wave may actually make it through to the opposite side of the barrier, allowing the particle to escape! This is shown schematically in Figure 15.3. The wave exponentially decays through the barrier, but once through, it resumes its normal wave (sinusoidal) pattern.

Penetration of the barrier instead of passing over the barrier is called *tunneling*. The effectiveness of tunneling depends on three factors. First is the difference between the energy of the particle and the barrier height. The smaller that difference is, the larger the tunneling efficiency will be. The second contributor is the barrier width. The narrower the barrier width, the greater the tunneling. The third factor is the mass of the object that might tunnel. The lower the mass, the greater the possibility of tunneling through a barrier. If one were to hunt for examples of tunneling, the ideal situations to seek would be light objects (like small atoms) contained within low, narrow barriers.

OK, all of this is fine, but is it real? Among the earliest identifications of tunneling was alpha particle decay. Uranium-238 is radioactive. Slowly, over time it disappears by emitting an alpha particle from its nucleus. The alpha particle is composed of two protons and two neutrons, or the helium-4 nucleus ($^4He^{2+}$). Uranium-238 ($^{238}U$) has 92 protons and 146 neutrons. After it expels the alpha particle, it becomes thorium-234, with a nucleus of 90 protons and 144 neutrons.

As often happens, the explanation for alpha decay was provided in two independent reports issued at about the same time by George Gamow and separately by Ronald Gurney and Edward Condon. They proposed a potential energy surface like the one shown in Figure 15.2. Conceptually, it may be easier to think of this surface as the one for the backwards reaction, capture of an alpha particle by $^{234}Th$, remembering that the forward reaction (alpha decay) and the backwards reaction (alpha particle capture) take place on the same surface. The alpha particle, being positively charged, will be repelled by the positive charge of the $^{234}Th$ nucleus. As the particle gets ever closer, the repulsion grows and grows, forming that upwardly curved barrier with decreasing separation. Most of the time, the alpha particle will simply scatter, or bounce, off the thorium nucleus. However, very rarely, if it hits at just the right angle with just enough energy, the alpha particle can overcome the repulsive force, penetrating the electric repulsive barrier, and enter the nucleus, where the strong force can join the alpha particle and $^{234}Th$ to make $^{238}U$.

Using this surface, the three physicists applied the then very new techniques of quantum mechanics and computed the wavefunction, the description of the alpha particle on this surface. The square of the wavefunction is the probability of finding the alpha particle at that point. The wavefunction is mostly within the well on the left, inside the nucleus. But some of that wavefunction leaks through the Coulomb barrier, and there is a nonzero wavefunction outside the nucleus. That's the means for the particle to tunnel through the barrier. The physicists' computations matched with the experimental observations of the rate of alpha decay, indicating the tunneling was observed.

Tunneling is not some esoteric phenomenon. Many home smoke detectors rely on the alpha decay of americium, a radioactive element just like uranium. A very small amount of americium is in the smoke detector, positioned a small distance away from an electronic device that registers when an alpha particle makes its way from the americium and through the air. If smoke is present, smoke particles will deflect the alpha particles from making their way to the detector, and this lack of signal will trigger the alarm. Tunneling is also involved in the operation of certain types of electronics, including tunnel diodes and tunnel junctions. Tunneling is a limitation to how small an integrated chip can be manufactured; if the circuits on the chip get too small and too close together, electrons can tunnel between sites instead of flowing in a managed and predictable way.

It is important to understand that our analogy to a tunnel has some real limitations. We drive a car, enter a tunnel, travel the length of the tunnel, and exit the opposite side. We have *traversed through the barrier*, knowing our position and velocity at every moment while we are in the tunnel. In the quantum world, this knowledge is forbidden. We only have knowledge of probabilities. When we square the wavefunction (that's the oscillating curve in Figure 15.3), we obtain the probability of finding the particle at a particular point. What the wavefunction of the alpha article tells us is the likelihood of finding the particle at some location or some region. Most likely, we will find the alpha particle within the uranium nucleus. But on some rare occasions, the alpha particle will be found *outside* the nucleus, and when we do find it, the electrostatic repulsion will push the alpha particle away from the resulting thorium nucleus.

Do not think of the alpha particle as having some trajectory starting in the nucleus and then traveling through the Coulomb barrier to the outside. The particle does not behave like our car passing through a mountain tunnel. It just suddenly appears outside the nucleus. All we can do is predict the probability that it will appear, not the actual moment that the particle passes through the barrier.

Some of the difficulty involved with fully grasping this idea is that we lack the proper language to talk about it. As Heisenberg noted, we live in a classical world where cars and trains travel through a tunnel and we can film and replay this trip. Our language was developed to describe our real experiences. The quantum world is outside our normal experience, and often we are just flailing around in an attempt to describe the beautiful mathematics using our inadequate language.

Tunneling is also observed in organic chemistry. The major examples involve rates, where kinetic isotope effects ($k_H/k_D$) greater than 7 have been

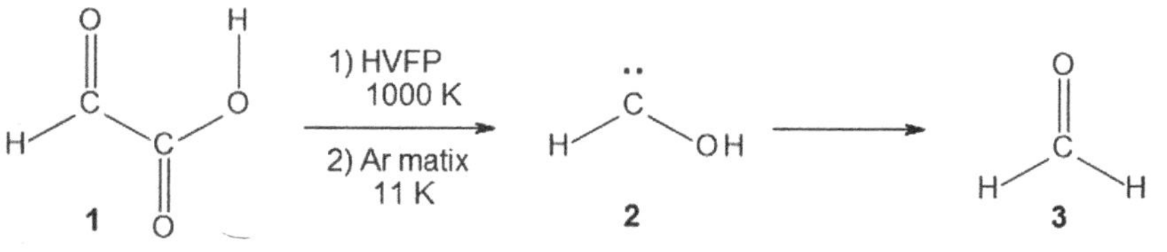

**Figure 15.4.** Decomposition of glyoxalic acid **1** within the HVFP experiment.

observed. These unusually large isotope effects are attributed to tunneling by the lighter $^1$H and a lack of tunneling by the heavier deuterium ($^2$H) isotope. Remember that mass difference is huge here: deuterium is twice the mass of hydrogen. And the larger the mass, the less likely the object is to tunnel. That's why we don't experience tunneling in our everyday life; there's just no way for our large body to tunnel to the head of the long line at the grocery store.

In the early twenty-first century, the German chemist Peter Schreiner published a series of papers identifying tunneling as the mechanism for some specialized reactions. These reactions involve the rearrangement of *carbenes*, a class of highly reactive organic molecules that have been used widely in organic synthesis. Carbenes are compounds in which a carbon atom makes two bonds and has a lone pair of electrons. Its electron count is therefore only six, making it a violation of the octet rule and providing the explanation for why they are so reactive. Typically, carbenes will insert between the atoms of a bond, meaning that they cause a bond to break and then attach to each of the resulting ends, making two new bonds. In another common reaction of carbenes, called *rearrangements*, the atoms in the carbene reattach in such a way as to complete the octet of all atoms.

Schreiner's first report was on the rearrangement of the carbene hydroxymethylene **2**. His intent was to make this elusive molecule and trap it at a very low temperature to characterize its properties. Schreiner suspected that the rearrangement to the stable compound **3** would be retarded at very low temperatures, allowing for a sufficient lifetime to measure a variety of its properties. The molecule was prepared by the reaction shown in Figure 15.4. The reactant **1** is glyoxalic acid. It is subjected to high-vacuum flash pyrolysis (HVFP). This technique exposes the molecule to intense heat, though for just a very short time. The compound is placed in a vial attached to a long tube that has been evacuated using a strong vacuum pump connected at the far end (see Figure 15.5.). This vacuum helps convert some of the liquid **1** into a gas, which starts to flow down the tube. The tube passes through a very intense oven, during which time **1** is heated to 1000 K, providing the energy needed to get the reaction to proceed to make **2**. The tube exits the oven and ends in

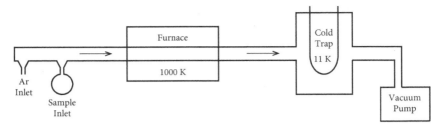

Figure 15.5. Schematic of the high-vacuum flash pyrolysis apparatus.

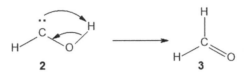

Figure 15.6. Mechanism for rearrangement of 2 into 3.

a chamber that contains a solid argon block at 11 K, a very, very cold surface. The exiting gas hits that cold argon surface and sticks to it. Whatever ends up surviving the process can then be examined on that argon surface.

Examination of the properties of the material captured in the pyrolysis of 1 confirmed the production of 2. However, over time, Schreiner discovered the disappearance of 2 and the appearance of formaldehyde 3. This finding was somewhat surprising at such a cold temperature. (11 K is very cold, just a little above absolute zero, the coldest possible temperature!) How could the rearrangement of 2 into 3 take place with so little thermal energy?

Accurate quantum mechanical computations of the potential energy surface suggest that no reaction of 2 should take place at all. There are two possible rearrangements from 2. The first involves the two hydrogens coming together to form $H_2$ and carbon monoxide, CO. The second rearrangement involves the hydrogen migrating from oxygen to carbon to form 3, which we can write using our arrow-pushing notation as shown in Figure 15.6. (Reaction mechanisms that involve arrows moving in a circle are the topic for the next chapter.) However, both reactions have very high activation barriers (Figure 15.7), much too large to be surmounted at this frigid temperature.

So if a reaction can't go over the barrier, the only option is to go through the barrier, that is, tunneling! Since only the light hydrogen has to move, and only from one neighboring atom to the other, that seems like a reasonable possibility. The natural test is to determine the effect of deuterium substitution; HC-O-D should rearrange much more slowly than does HC-O-H since the

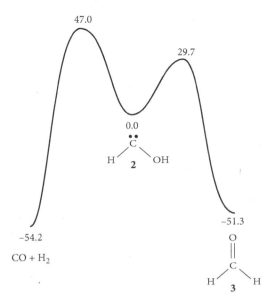

**Figure 15.7.** Potential energy surface of the transformation of **2** into **3** or into CO + H$_2$. Energies are in kcal mol$^{-1}$.

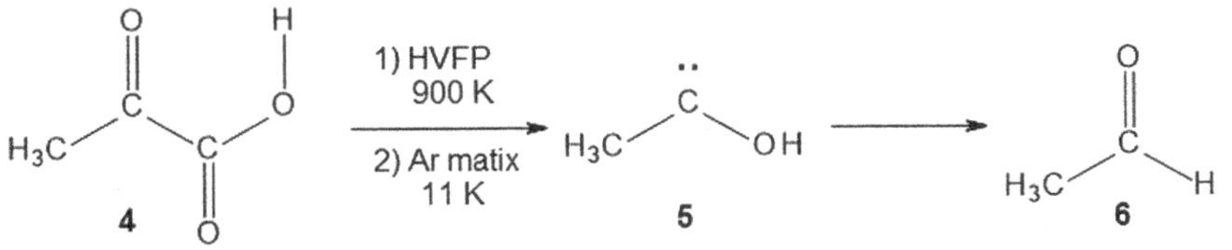

**Figure 15.8.** Decomposition of 2-oxopropanoic acid **4** within the HVFP experiment.

extra mass of deuterium should significantly dampen tunneling. Schreiner made this deuterated analogue, and it doesn't react at all, even over multiple days. Here's a reaction that takes place *because* of tunneling.

An even more interesting situation takes place with hydroxyethylene **5**. It, too, is made in the HVFP reactor, via the reaction shown in Figure 15.8. The acid **4** is heated to 900 K, and the product gas is collected at 11 K. The carbene **5** is confirmed by examining its properties. Again, over the course of a couple of hours, **5** disappears and acetaldehyde **6** appears.

As with carbene **2**, the rearrangement of **5** to **6** involves a light hydrogen atom migrating from oxygen to the neighboring carbon, which is just a short hop away. Again, that's a prime setup for a tunneling reaction. The deuterated analogue of **5**, CH$_3$-C-O-D, was made, but it does not rearrange at all at this

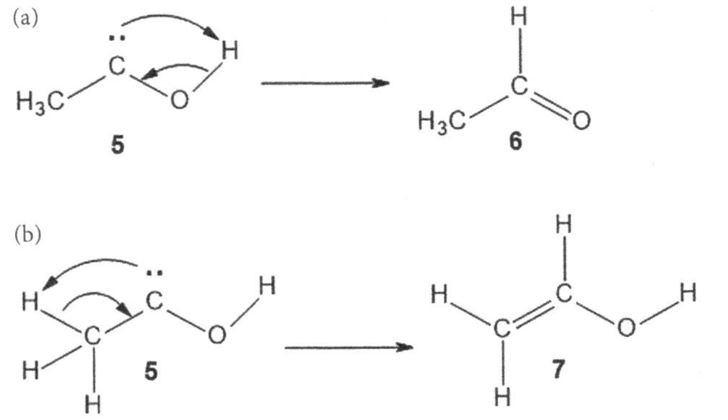

**Figure 15.9.** Mechanism of the rearrangement of carbene **5** into (a) acetaldehyde **6** or (b) vinyl alcohol **7**.

cold temperature. Again, this strongly suggests that the hydrogen migrates by tunneling through the barrier.

Carbene **5** has two reasonable rearrangement pathways, one to form acetaldehyde **6**, which is observed, and one to form vinyl alcohol **7**, which is not observed. The mechanisms for these two rearrangements are shown in Figure 15.9. Both involve a hydrogen migrating from a neighboring atom to the carbene center. To get from **5** to **6**, the hydrogen on oxygen moves to the neighboring carbon, also making the C–O double bond. The reaction of **5** to **7** has a hydrogen from the terminal carbon move to the carbene center while forming the C–C double bond.

The experimental result, only **6** is observed, seems a bit unusual given the computed potential energy surface for these two rearrangements shown in Figure 15.10. The barrier to form **7** is lower than the barrier to form **6**, so all things being equal, tunneling should preferentially proceed through that lower barrier. Tunneling rates depend on the height of the barrier. But they also depend on the width of the barrier. The distance the hydrogen has to move to transform **5** into **6** is significantly shorter than the path to form **7**. The other two hydrogens attached to the end carbon of **5** also have to move to make the planar arrangement about the double bond in **7**. So, that barrier to make **7** is much wider than the barrier to make **6**. As Schreiner and his co-authors cogently argue, "barrier width trumps barrier height."

What "controls" this reaction of **5** going to **6**? It's not thermodynamic control; it's way too cold for that to matter! It's not kinetic control because the lower barrier leads to form the unobserved product **7**. It's a third means, what Schreiner has called "tunneling control", that dictates which product is made.

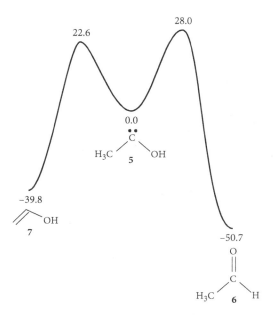

**Figure 15.10.** Potential energy surface of the transformation of **5** into **6** or into **7**. Energies are in kcal mol⁻¹.

I'm sure that if one had polled physical organic chemists in the year 2000 about the possibility of another option existing beside kinetic control and thermodynamic control, no one would have responded in the affirmative. This was well-worn ground and well understood for more than half a century. These concepts are integrated into every introductory organic chemistry textbooks. What surprises could be out there? It was beyond the imagination to consider a new control process. The lesson here is that we must remain vigilant in our experimental quests. The discovery of tunneling control demonstrates that new treasures await us still; one just needs an inquisitive interest and an open mind.

# 16
# Aromaticity

The eminent British scientist Michael Faraday, a jack-of-multiple-trades scientist, made important contributions to both physics and chemistry. It was his work in these two disciplines that was the inspiration for naming the building at Northern Illinois University where I began my academic career. The building bore his name, even though he never visited DeKalb, Illinois, and quite fittingly, it housed both the chemistry and physics departments.

Although Faraday is probably best known for his work on magnetism, this chapter centers on one of his accomplishments in chemistry. In one of his many experiments, Faraday focused on the process to make the "portable gas" used for lamps. The gas was produced by rapidly heating whale oil or fish oil and then pressurizing the resulting gas. Faraday collected a liquid by-product of the pressurization; that sample can still be seen today in the Faraday Museum in London. Analysis of that liquid identified a new substance, which was later called benzene. Subsequently, benzene was also found in petroleum and coal.

Faraday showed that benzene contained just carbon and hydrogen, and soon thereafter, it was found to have the molecular formula $C_6H_6$. Many compounds known in the late 1800s contained only carbon and hydrogen, but the ratio was typically much more in favor of hydrogen. In fact, many compounds have more than twice as many hydrogen atoms as carbon atoms. Organic chemists refer to a *degree of unsaturation* to quantify this missing amount of hydrogen that could be present. The early problem of the structure of benzene centered on how to have each carbon atom participate in four bonds with so few hydrogen atoms present.

The German chemist August Kekulè's first attempt at a structure for benzene was a six-membered ring with alternating single and double bonds 1. Every carbon makes four bonds. The problem with 1 is that it fails to explain the structures of the disubstituted benzenes. If we attach two groups to 1, we can predict that there will be four different isomers 1a–1d (Figure 16.1). Carefully note that 1a and 1b differ by having the two neighboring carbons with the substituents connected by a single bond in the former and a double bond in the latter. However, only three isomers are actually observed.

Thinking Like a Physical Organic Chemist. Steven M. Bachrach, Oxford University Press. © Oxford University Press 2023.
DOI: 10.1093/oso/9780197640371.003.0016

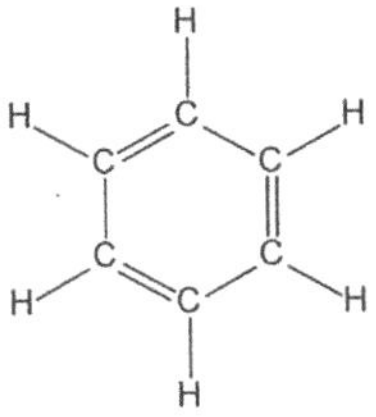

**Figure 16.1.** Alternating double-single bond representation of benzene and the disubstituted benzene isomers.

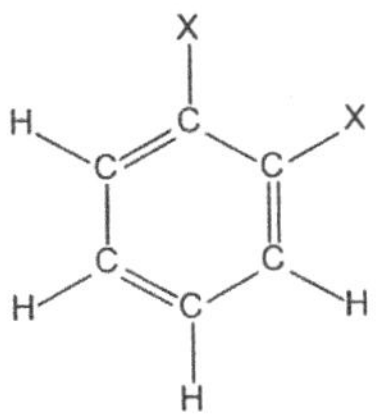

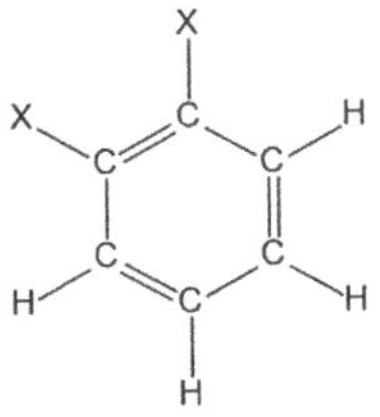

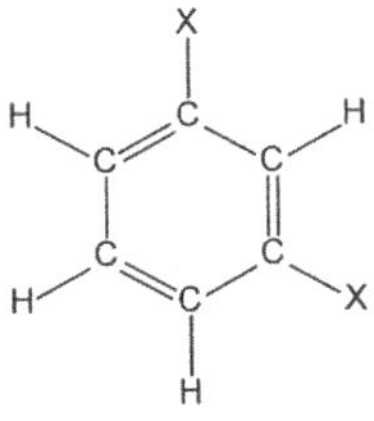

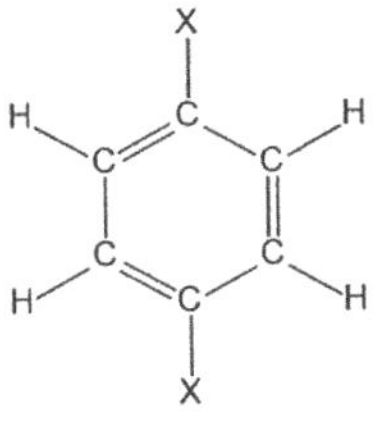

In 1872, Kekulè offered a new model for benzene. His model is associated with perhaps the most famous story in the history of organic chemistry. Kekulè professed to have had a dream in which a snake formed a circle and bit its own tail. In his own words:

My mind's eye, sharpened by repeated visions of a similar sort, now distinguished larger structures of varying forms. Long rows frequently close together, all, in movement, winding and turning like serpents. And see! What was that? One of the serpents seized its own tail and the form whirled mockingly before my eyes. I came awake like a flash of lightning. This time also I spent the remainder of the night working out the consequences of the hypothesis. (*Berichte der Deutschen Chemischen Gesellschaft*, 1890, pp. 1305)

That image in his dream led Kekulè to propose that benzene arose from two structures **2a** and **2b** oscillating back and forth so quickly that neither can be identified; only the average of the two structures is seen (Figure 16.2). (This story may have been apocryphal, for in the many years after the event, Kekulè regaled different audiences with conflicting accounts of his discovery. Nonetheless, this self-consuming-snake tale remains a staple of many organic chemistry textbooks.)

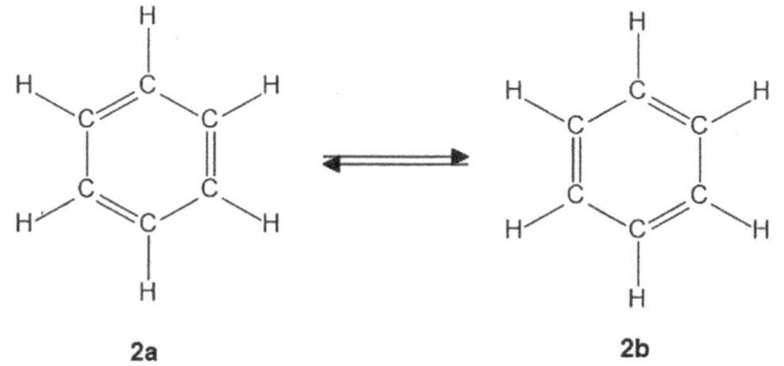

**Figure 16.2.** Kekulè's equilibrium model of benzene.

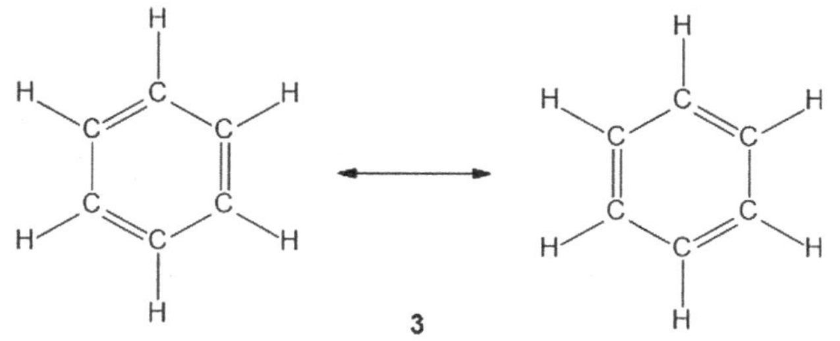

**Figure 16.3.** Resonance structure description of benzene.

Based on what I have already presented in this book, you are likely to know what happens next in this story. **2a** and **2b** don't look like different physical structures; rather, they are different *resonance structures*. Linus Pauling's resonance theory nicely explains the structure of benzene. In order to properly describe benzene, two resonance structures are required, as indicated by **3** in Figure 16.3. This notation means that each structure contributes to the proper description of the molecule. The molecule is the sum of 50% of the left structure and 50% of the right structure. Benzene is not bouncing back and forth between **2a** and **2b**, having the C–C bonds oscillate between a short length (the double bond) and a long length (the single bond). Rather, each carbon–carbon bond has a bond order of one and a half. One can think of this as analogous to the unicorn. This legendary creature is part rhinoceros and part horse all the time, not some of the time a rhino and some of the time a horse, and the time average is the unicorn. It is a unicorn all the time.

We can think of the six electrons in the three double bonds as circulating around the entire ring. That leads to an alternative pictorial representation having a hexagon with a circle in the middle as in **4** (Figure 16.4). The

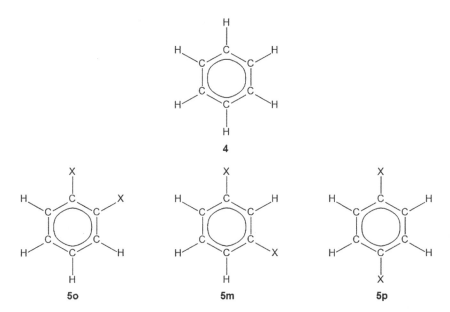

**Figure 16.4.** Resonance description of benzene and disubstituted benzene isomers.

representations **3** and **4** are equivalent and are just different ways to show the distribution of the 6 electrons around the benzene ring.

This circle-in-the-hexagon representation makes it obvious that all six carbons in benzene are identical and that all C–C bonds are identical. That equivalence was confirmed with the determination of the structure of benzene that indicates a perfect hexagonal structure with six equivalent bond lengths. The equivalence of the six carbon atoms also helps explain why there are three and only three isomers of disubstituted benzene: the isomer where the two substituents are on adjacent (called *ortho)* carbon atoms **5o**, on carbons one removed (called meta) **5m**, and on the carbons opposite each other (called *para*) **5p** (see Figure 16.4).

Each individual resonance structure of **3** shows three double bonds. It might be reasonable, then, to expect benzene to react like an alkene. I discussed addition reactions in Chapter 14, and addition across the implied double bonds in benzene can be experimentally tested. An addition reaction that I did not discuss in Chapter 14 is the reaction of $Br_2$ with an alkene. An example of this reaction is the addition of $Br_2$ to cyclohexene **6** to give 1,2-dibromocyclohexane **7**. As expected, one addend, a bromine atom, attaches to each carbon of the double bond (see Figure 16.5a). There is interesting stereochemistry here too: note that the bromine atoms add from opposite sides. We'll leave that alone here, though physical organic chemists have developed a very interesting mechanism to account for this stereoselectivity.

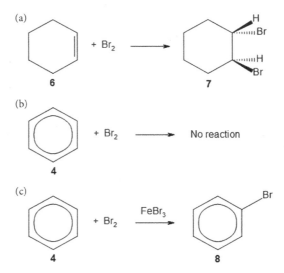

**Figure 16.5.** (a) Bromination of cyclobutene, (b) the failed bromination of benzene under the same condition, and (c) bromination of benzene with a catalyst.

Given this result, one might expect that $Br_2$ would add across the double bonds in benzene, maybe as many as three times, to have one bromine attached to each carbon. However, if we mix benzene and $Br_2$ we get no reaction at all (Figure 16.5b); starting material is recovered. If we provide a catalyst, which could be $FeBr_3$ (as shown in Figure 16.5c), a reaction does occur, and the product is bromobenzene **8**. *Only one bromine attaches to benzene.* This is not an addition reaction, but rather a substitution reaction! A hydrogen atom is replaced by a bromine atom. I will discuss this mechanism later in this chapter.

Next I present the results of another addition reaction, this time adding $H_2$ across the C–C double bond. This reaction requires a catalyst, whether the reactant is an alkene (like cyclohexene **6**) or benzene. The typical catalyst is a platinum surface, but other metals also catalyze this reaction. High pressure is often necessary too, making this reaction very dangerous if not performed with the proper equipment; $H_2$ mixed with oxygen powers rockets, like the one used to launch the Space Shuttle.

The addition of $H_2$ across the double bond of cyclohexene **6** makes cyclohexane **9**. The reaction is exothermic, releasing 120 kJ of energy (Figure 16.6). You might expect that the addition of $H_2$ to 1,3-cyclohexadiene **10** to give cyclohexane would release twice the amount of energy as the addition reaction with cyclohexene, since there are twice the number of double bonds. The reaction of **10** is exothermic, but it releases 232 kJ, slightly less than twice that in

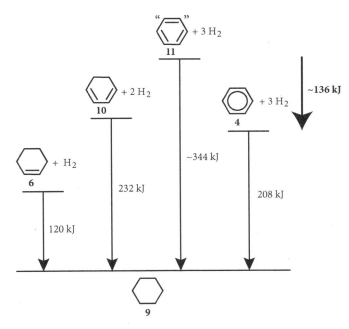

**Figure 16.6.** Reaction energy diagram to propose the aromatic stabilization energy of benzene.

the addition reaction of **6**. This shortfall can be attributed to some stability of the conjugated diene.

Imagine if benzene were described by just one of the Kekulè structures, with alternating single and double bonds around the ring. This turns off the resonance, and we can describe it as **11**. The quotation marks in Figure 16.6 indicate that this is an hypothetical structure, or what Einstein would call a *gedankenexperiment,* a thought experiment. We might then infer that **11** has three double bonds that are conjugated. We get 120 kJ for adding $H_2$ across that first double bond and 112 kJ for adding across that second bond. Let's assume we get another 112 kJ for adding across the third double bond, for a total of 344 kJ released by complete hydrogenation.

Next, let's fully hydrogenate benzene **4** in a true laboratory experiment. The reaction is again exothermic, releasing 208 kJ (Figure 16.6). That is significantly less than our estimated value of 344 kJ. The bold arrow on the right of Figure 16.6 indicates that benzene is about 136 kJ *more stable* than what we anticipated. That's a sizable stabilization energy, suggesting that benzene will be less reactive than expected. In fact, this already hints at some explanation of the substitution reaction in Figure 16.5c. Note that the benzene core remains unchanged in the reaction and that only a hydrogen is replaced by a bromine; the bonds within the ring, between the carbon atoms, are unchanged.

What is the root of the stabilization energy of benzene? An early explanation suggested there was something special about alternating single and double bonds (conjugation) around a ring. However, the fact that cyclobutadiene **12** has still not been isolated as a freestanding molecule calls that suggestion into doubt. In the 1920s, this suggestion was modified to "there is something special about having 6 electrons in a conjugated ring." However, no explanation was provided as to why 6 electrons represented such a magical number that imbued stability. To be fair, at that time, chemistry seemed to have a few magic numbers. Helium with its 2 electrons, neon with 10 electrons, argon with 18 electrons, and xenon with 56 electrons—all of these are inert. The elements near these inert gases on the periodic table gained some stability when they acquired the same number of electrons. Magic numbers were just in the air.

**12**

In the late 1920s and early 1930s, quantum mechanics again came to the rescue, this time with the pioneering work of the German physicist Erich Hückel. Perhaps influenced by his brother Walter, an accomplished organic chemist (though Hückel later commented that it was probably the German physicist Friedrich Hund who guided him), Hückel applied quantum mechanics to benzene and similar molecules. Hückel's major prediction is now known as *Hückel's rule*: a cyclic conjugated π-system will be especially stable if the number of electrons in the π-system is equal to $4n + 2$, where $n$ is an integer. Benzene **4** has 6 π electrons from the three double bonds; that satisfies Hückel's rule because when $n$ is 1, $4(1) + 2 = 6$. This rule also predicts that molecules with 10, 14, 18 electrons and so on will also be especially stable. For example, naphthalene **13** has 10 π electrons spread over two rings. Each of the resonance structures of **13** has alternating single and double bonds, so that the π electrons are spread over the whole molecule. It is planar, the C–C distances are almost identical, and naphthalene undergoes substitution reactions. **13** is certainly aromatic.

**13**

Hückel's rule also predicts that molecules that don't have $4n + 2$ π electrons will not be stabilized. In fact, Hückel goes even further. His second rule is that molecules that have $4n$ π electrons, again with $n$ as an integer, will be destabilized and are unlikely to exist. Cycobutadiene **12** has 4 electrons, and that satisfies the $4n$ electron rule. As I noted before, **12** remains unknown.

Molecules like benzene and naphthalene are grouped together under the concept of *aromaticity*. This term derives from the pleasant smell of benzene and its cousins, though the compounds in perfumes and fruits are often not of this class. Aromatic compounds are best categorized by their properties:

1. They tend to be planar, made up of one or more rings.
2. The bonds around the ring are of equal length.
3. They observe Hückel's rule, having $4n + 2$ π electrons.
4. They express certain magnetic properties (though I will not discuss these properties in this book).
5. They prefer substitution reactions to addition reactions.

Application of Hückel's rule suggests that charged molecules might be aromatic. For example, cyclopentadiene **14** is not an aromatic compound since it does not have π electrons delocalized around a ring; the molecule has no resonance as it can be properly described with just a single structure diagram. However, if it were to act as an acid and give up an H$^+$ (as in the reaction shown in Figure 16.7), it would form the cyclopentadienyl anion **15** that requires five resonance restructures to delocalize the charge (and the six electrons) around the ring. This anion is planar and the C–C distance are identical. It is decidedly aromatic!

Compounds with $4n$ electrons are called *antiaromatic*, largely because they are so unstable. Cyclooctatetraene **16** has eight π electrons, suggesting that it should be antiaromatic, just like cyclobutadiene **12** which has four π electrons.

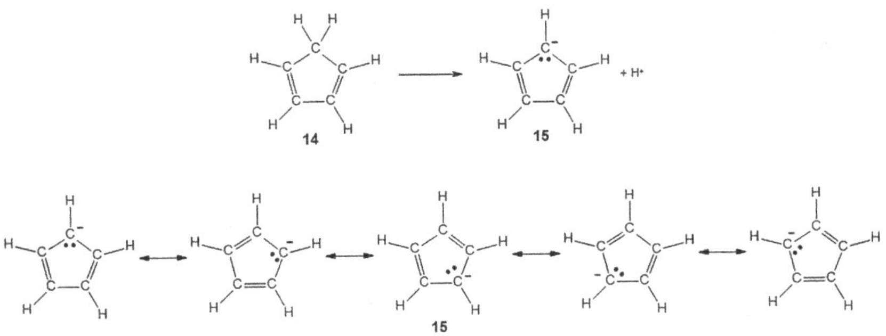

**Figure 16.7.** Enhanced acidity of cyclopentadiene **14** explained through the resonance of the anion **15**.

However, unlike **12**, **16** has been prepared and can be stored indefinitely in a bottle. It will readily undergo addition reactions like any ordinary alkene. If **16** really is antiaromatic, and if we can prepare it, at best it should be highly reactive and is unlikely to hang around in a jar.

The key to understanding why we can have a bottle full of **16** is its structure. Remember that aromaticity is associated with the delocalization of the π electrons around the ring. This delocalization is best if the molecule is planar so that all of the p-orbitals that make up the π bond are aligned in parallel. The structure of **16** is shown in Figure 16.8. It is not planar but rather adopts a tub shape. The bonds are localized, meaning distinct alternation of short–long–short bonds as you go around the ring. **16** avoids a planar structure in order to evade the antiaromatic character associated with a planar *4n* electron ring. **16** is neither aromatic nor antiaromatic; it's just an ordinary alkene with multiple double bonds.

A very interesting thing happens when we react **16** with a metal like potassium. Potassium is prone to give up an electron, and if each of two potassium atoms releases an electron and a cyclooctatetraene molecule (**16**) grabs them both, the resulting dianion **17** will have ten π electrons: it started with eight electrons and adds two more. Hückel's rule indicates that a delocalized cyclic π system with ten electrons should be aromatic. Well, the structure of the diananion of **17** *is* planar with all eight C–C bonds of identical length!

Cyclooctatetraene **16** provides an excellent proof-of-concept for aromaticity and antiaromaticity. Antiaromaticity is an unfavorable circumstance, and **16** adopts the tub shape to minimize the destabilizing effect of antiaromaticity. But drop in two additional electrons into the π system and the molecule transforms, becoming planar and expressing all of the attributes one associates with the stabilizing nature of aromaticity.

The main value of the concept of aromaticity is that it allows us to aggregate hundreds of thousands of molecules into one tidy category. Aromatic

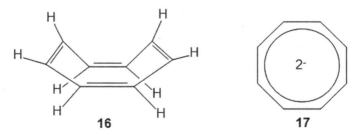

**Figure 16.8.** Tub-shape of cyclooctatetraene **16** and the planar structure of its dianion **17**.

compounds present a range of variation: single rings, multiple rings, neutral or charged, some with atoms other than carbon in the $\pi$ backbone. Aromatic compounds are components of DNA and RNA, pharmaceuticals, poisons, and building blocks for plastics like Kevlar. Some are quite carcinogenic. The unifying concept of aromaticity helps us understand this range of properties and applications.

Let's talk about the mechanism for the substitution reaction of aromatic compounds, like the one shown in Figure 16.5c. The aromatic ring is electron rich, with those six $\pi$ electrons delocalized over the ring. A reactant that carries some positive charge will be attracted to that electron-rich ring. A positively charged group, or any group that is attracted to negative charge, behaves as an *electrophile*, a group that "loves" electrons, the carrier of negative charge. So, this reaction can be classified as *electrophilic aromatic substitution*.

The reaction of benzene and bromine ($Br_2$) (the reaction shown in Figure 16.5b) does not occur because bromine is not a strong enough electrophile. But when mixed with $FeBr_3$, it forms $Br-Br^+-Fe^-Br_3$, with the central bromine atom having some positive charge (see Figure 16.9). That makes the bromine a better electrophile, and so one bromine atom adds to one of the carbons of benzene, forming the intermediate cation **18** (Figure 16.9). This cation is stabilized by the three resonance structures that delocalize the positive charge about the ring.

If a normal addition reaction were to take place, the second bromide anion would attach to the carbocation center, forming the dibromobenzene **20** (Figure 16.10). Note that this product has destroyed the aromaticity of the ring; there are only four $\pi$ electrons, and no longer are there alternating single and double bonds about the ring. Loss of aromaticity means loss of the accompanying stabilization energy, making this addition step quite costly in terms of energy—and thus unlikely to occur.

Instead, the intermediate **18** acts as an acid and gives up an $H^+$, which ends up in HBr (Figure 16.9). The loss of the $H^+$ reforms the third double bond, reestablishing the aromatic ring. The net overall change is simply a substitution of a hydrogen with a bromine atom. The aromatic ring remains unchanged, reflecting its great stability. This stability differentiates the reactivity of aromatic molecules from the alkenes.

Like all scientific models, aromaticity has its limitations. Exploration of these limits helps us understand the model in more detail and guides us to new discoveries. I will next present examination of these limits by looking at the reactivity of some aromatics, followed by an exploration of the geometric attributes of aromaticity, namely, having a planar structure and the equivalence of the C–C bonds about the ring.

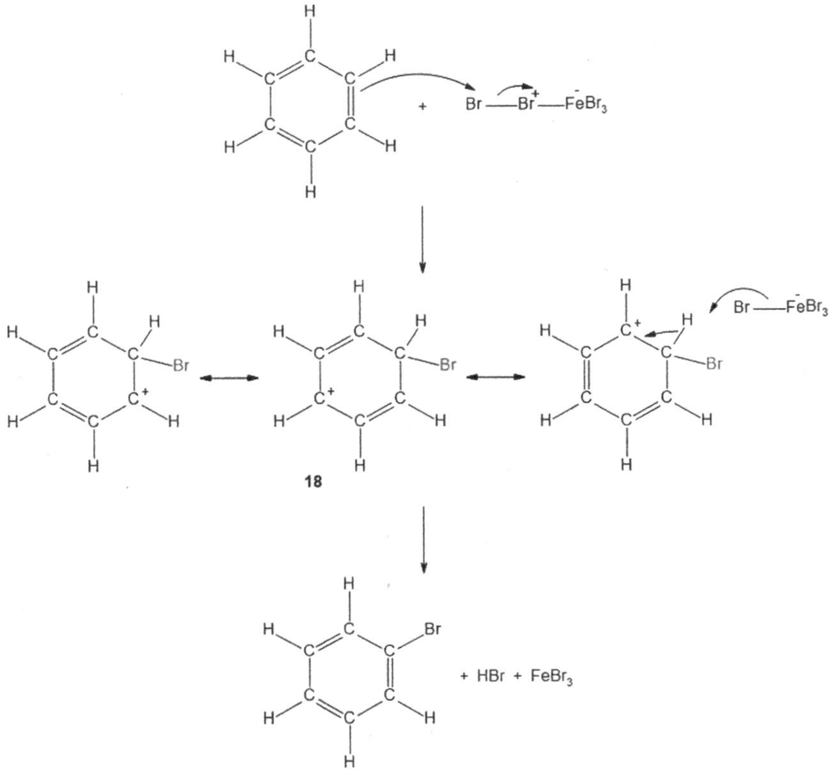


18

19

**Figure 16.9.** Mechanism for the bromination of benzene.

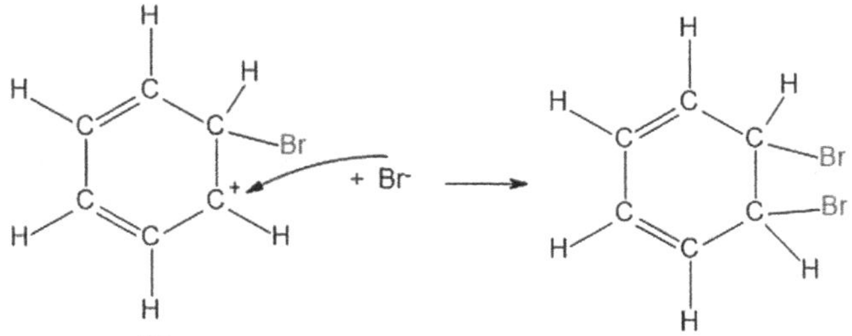


18

20

**Figure 16.10.** Loss of aromaticity in the addition of two bromine atoms to benzene.

The majority of aromatic compounds will undergo substitution reactions. For example, substituted benzenes such as **21** and **23** (Figure 16.11) will readily undergo substitution reactions. The examples shown in this figure are nitration reactions. Interestingly, **21** and **23** add substituents with regioselectivity. **21** will react such that the added group ends up in either the *ortho* or *para* position, but virtually no *meta* product is observed. On the other hand, **23** undergoes substitution such that only the *meta* product is produced. Physical organic chemists have devised a beautiful model to explain this selectivity, making extensive use of resonance structures. It's a model that defines how an existing substituent directs where the next substituent will end up. When I teach organic chemistry, my least favorite lecture is the one where I present this explanation because I end up drawing dozens of resonance structures on the whiteboard and it is simply very tiring! I'll skip the details here.

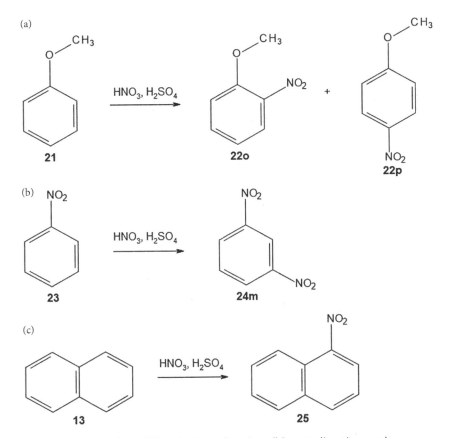

**Figure 16.11.** Examples of (a) ortho/para direction, (b) meta direction, and (c) polycyclic aromatic direction.

This model of substituent directing effects also explains why electrophilic substitution of naphthalene **13** yields just a single product, **25** (Figure 16.11c). If you are interested in accessing the other possible isomer, that preparation is much more complicated and requires multiple steps.

Which brings us to the reaction of phenanthrene **26** (Figure 16.12a). Phenanthrene appears to be a normal aromatic compound. It has 14 π electrons, so it satisfies Hückel's rule: $4(3)+2 = 14$. It is planar, and while the C–C distances are not identical, they are very close in length to each other. Its magnetic properties are consistent with other aromatic compounds. However, phenanthrene will react with $Br_2$ without the need of a catalyst, and it undergoes an addition reaction to give the dibromo compound **27** (Figure 16.12a). That top double bond is lost and seems to react just like an ordinary alkene.

The explanation of this behavior is wrapped up in the degree of aromatic stabilization. On a per carbon basis or a per ring basis, benzene is more stabilized by aromaticity than is naphthalene **13**, which in turn is more stabilized than phenanthrene **26**. Its lessened stabilization energy means that **26** will behave chemically more like an alkene and participate in addition reactions. Note also that the product of this addition, **27**, has two aromatic rings isolated from each other, each with the stabilization of benzene. This is a highly stabilized molecule. Similar behavior is seen in the reaction of another three-ring

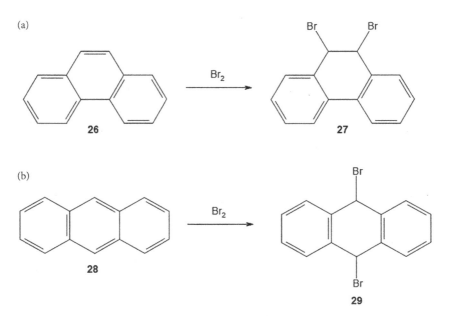

**Figure 16.12.** Two examples of where aromatic compounds will preferentially undergo addition reactions over substitution reactions,

aromatic, anthracene **28**, which also adds $Br_2$ (Figure 16.12b). It too results in a product with two benzene-like rings. The message here is that not *all* aromatic compounds will favor substitution over addition reactions.

Next up is the notion that aromatic compounds are planar. All of the example molecules I have presented so far are planar. Coronene **30** is a perfectly flat molecule, with a central benzene ring surrounded by six more rings. This is a fragment of graphite, the material found in pencil lead. These types of graphite fragments are called *graphenes* and are of interest for their electrical and magnetic properties.

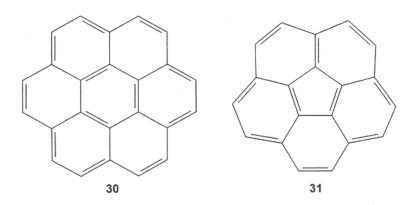

**30**                                          **31**

If the central ring of coronene is replaced with a five-membered ring, we have corannulene **31**. That seemingly small change actually puts significant strain on the angles at the interior carbons, resulting in **31** having a bowl shape (Figure 16.13). The magnetic properties, bond lengths, and preference for substitution reactions nonetheless suggest that this nonplanar molecule manifests in some aromaticity.

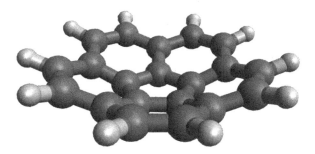

**Figure 16.13.** Bowl structure of corannulene **31**.

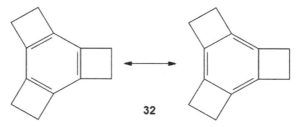

**Figure 16.14.** Resonance structure of **32**.

The last aspect of aromaticity to discuss here is the notion of equivalent bond lengths about the aromatic ring. All six bonds in benzene are of identical length. They are almost of identical length in naphthalene, anthracene, and phenanthrene as well. William Mills and Ivor Nixon wondered whether, by appending small rings to benzene, the strain from those added rings might localize the bonds, leading to real alternation of bond length about the ring.

For example, **32** has three four-membered rings fused to benzene. An ordinary four-membered ring is strained because of the narrowed angles required to form the ring; the angles about a carbon making four distinct bonds are typically about 110°, not the approximately 90° needed in this small ring. Putting a double bond into a four-membered ring increases the strain substantially. Recall that the angles about a double bond prefer to be 120°. Mills and Nixon thought that the left resonance structure of **32** shown in Figure 16.14 would be less likely to participate in describing the molecule because of the strain of having the double bond within the four-membered ring. If the right resonance structure dominates the description of the molecule, then there should be distinct alternation of the bond lengths about the six-membered ring, with the longer bond being the one shared by the four- and six-membered rings. In fact, the difference between the bond lengths within the six-membered ring is very small—about what is observed in phenanthrene and anthracene. Apparently, aromaticity is so stabilizing that it overcomes the strain of those three fused small rings.

Can we torture a poor benzene ring a bit more? How about putting an additional double bond into each four-membered ring of **32**, making **33**? In addition to adding much more strain to the molecule, the resonance structure on the left in Figure 16.15 has three cyclobutadiene fragments, each of which, having 4 π electrons, is antiaromatic. That left resonance structure should therefore be much less stable than the resonance structure on the right. That implies that **33** will be dominated by the right resonance structure shown in Figure 16.15. That right resonance structure means a longer C–C bond where the bond is shared by two rings and a shorter bond in between. All of those unfavorable cyclobutadiene

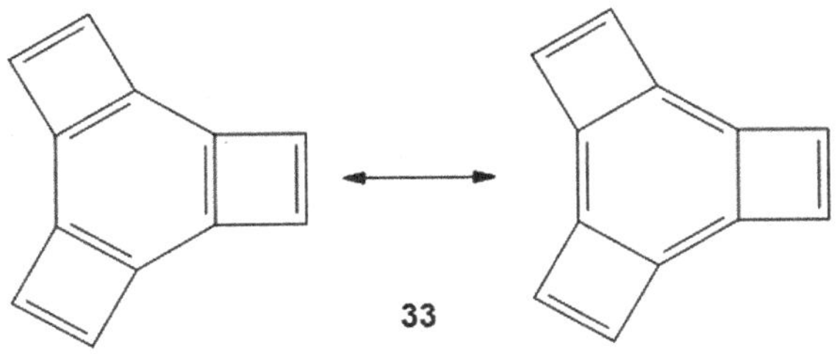

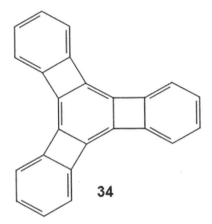

**Figure 16.15.** resonance structure of **33**.

**Figure 16.16.** Structure of starphenylene **34**.

fragments makes **33** difficult to synthesize, and it is still unknown. However, quantum mechanical computations do suggest a significant bond length distance difference around the six-membered ring of **33**.

The American chemist Peter Vollhardt devised a very clever analogue of **33**. The problems involved in preparing **33** are those antiaromatic four-membered rings. Instead of that second double bond, Vollhardt posited that some reduction in the double-bond character might be a better synthetic target. He then successfully prepared the analogue **34**, called starphenylene, shown in Figure 16.16. Each terminal benzene ring shares a 1.5 bond with the four-membered ring, so the four-membered ring should avoid most of the potential antiaromatic character. In fact, the resonance structure drawn in Figure 16.16 entirely avoids the cyclobutadiene ring.

Now **34** has some very interesting properties. The bond distances in the central six-membered ring are spot on for typical single bonds and double bonds. They certainly reflect double-bond localization to the positions between the fused bonds. Perhaps even more intriguing is that the energy released in adding three molecules of $H_2$ (hydrogenation) across the three double bonds of that central ring is equivalent to three times the energy of adding $H_2$ across the double bond of cyclohexene (the far left reaction in Figure 16.6). This equivalence in energy indicates that there is *no stabilization due to aromaticity in that central ring of 34*. Vollhardt seems to have prepared that *gedanken-ring*, the hypothetical ring, "cyclohexatriene," the fictional compound 11!

What is the upshot of all this probing of the edges of aromaticity? Aromaticity isn't black or white; molecules are not either aromatic or not aromatic, antiaromatic or not antiaromatic. Once again, what we see in nature is a continuum, molecules that express a range of aromatic properties and behaviors. What is unusual with aromaticity is that there is really only one molecule that fully expresses the properties ascribed to this concept—and that is benzene. All other "aromatic" molecules will defy the properties to some extent: the molecule may be nonplanar; or the bonds are not of equivalent distance; or the stabilization energy is reduced; or it undergoes additional reactions; or some combination of these.

It is fair to wonder about the value of creating a group that has just one true member with a whole host of members with varying degrees of affiliation. For organic chemists, the concept of aromaticity still has enormous value. We can group together millions of molecules and know that there is some commonality, that we can expect every aromatic molecule to display some set of properties. We can draw upon our arsenal of reactions that aromatic molecules will undergo, expecting as well that a whole slew of reactions will not complicate matters. We can anticipate certain electrical and magnetic properties and build on these expectations to design new molecules.

Aromaticity seems to be rather robust. Significant strains can be placed on these rings, and they retain their aromatic character. Note that 32 and even its analogue where the four-membered rings are replaced with the even more strained three-membered rings (35) still have nearly identical bond lengths about the ring and sizable stabilization energy. Chemists have examined planar aromatics that are forced to bend out of plane, like 36, and they too retain their stabilization energy.

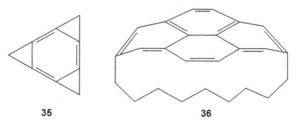

35                                    36

Perhaps most importantly, we will not be shocked when some of these expectations fail us. This is not just a case that some organizing principle is better than none at all. Aromaticity is an extremely valuable concept that has guided chemists for over a century. It has served us well in understanding important breakthroughs in science, maybe none more critical than in decoding the structure of DNA. The concept of aromaticity greatly enabled the scientists who determined the structure and function of DNA. The familiar double helix of DNA is held together by flat base pairs, each of which is an aromatic molecule. These base pairs stack upon each other, twirling about like a spiral staircase. These flat, aromatic molecules attract the flat base pair above and below it, helping to stabilize the molecule. Some carcinogens are flat aromatic molecules. They owe their deleterious physiological behavior to the fact that they fit in between the stacked base pairs, like a piece of paper sliding in between pages of a book.

Aromaticity is a model, an organizing guide that helps us manage a large grouping of molecules. Like all models within science, aromaticity provides a framework for considering commonalities of behaviors and properties. It allows us to predict these behaviors and properties prior to synthesis, and it helps guide us as to what molecules may have the right properties we seek. But like all models in science, aromaticity begins to fail as we move to the extremes of its coverage. Continuous probing of these extremities helps us refine the model, improving its robustness, and opening vistas onto new areas to explore and define.

Aromaticity also played a role in leading chemists to the pinnacle of physical organic thought, its crowning achievement, which is the subject of our next chapter.

# 17

# Pericyclic Reactions

To set the stage for the reactions presented in this chapter, first I need to discuss another type of molecular isomerism. Remember that isomers are molecules that have some things in common and some properties that are different. For example, I previously discussed enantiomers, molecules that are mirror images but are not superimposable on each other, just like your left and right hands are identical except that they are mirror images.

Geometric isomers are molecules that have the same bonds, but when the bonds are connected in three-dimensional space, the molecules differ by having some groups arranged on the same side, but in the other arrangement, the groups are on opposite side. When the groups are on the same side, we call that arrangement *cis*. A simple example might be a serving tray (Figure 17.1a) where the two handles are on the same face of the tray. We would identify the arrangement of these handles as *cis*. One might imagine making a fatal error in building such a tray and placing one handle on each face. That arrangement, with the handles on opposite sides, we call *trans*, and obviously the *trans* serving tray will have little practical value. On the other hand, the bookstand shown in Figure 17.1b has a base, with a prop on one end to tilt the base, and at the opposite end, on the other side is a prop for the book to rest upon.

Two types of common organic molecules display this *cis/trans* isomerism. Cyclic compounds can have groups attached on the same face (*cis*) or on opposite faces (*trans*). For example, see the two substituted cyclobutanes depicted in Figure 17.2. The molecules have identical bonding: a four-membered ring made up of C–C bonds, and two C–Cl bonds on adjacent carbon atoms. However, one has the two chlorine atoms on the same side, and the other has them on opposite sides. This manifests in real differences, such as their boiling points, densities, and reactivity.

Importantly, it is not possible to interconvert the *cis* and *trans* geometric isomers of the ring system, like those shown in Figure 17.2 without breaking bonds. The interconversion can only be done by cleaving the C–C bond between the two chlorine atoms, rotating one of the ends, and then reattaching.

*Thinking Like a Physical Organic Chemist.* Steven M. Bachrach, Oxford University Press. © Oxford University Press 2023.
DOI: 10.1093/oso/9780197640371.003.0017

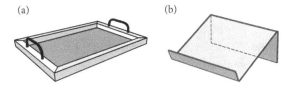

**Figure 17.1.** Models of a *cis* arrangement (a) a serving tray and a *trans* arrangement (b) a bookstand.

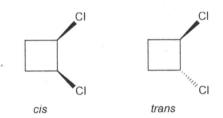

**Figure 17.2.** *Cis* and *trans* 1,2-dichlorocyclobutane.

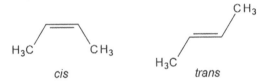

**Figure 17.3.** *Cis* and *trans* 2-butene.

That process takes significant energy, and so these two molecules can be made, each placed in separate jars, and, if left alone, they will never change into the other isomer.

A second type of geometric isomer may occur in alkenes. This isomerism arises from the fact that rotation about the double bond is extremely difficult; for all intents and purposes, the C = C is locked. That means that substituents attached to each carbon of the double bond are frozen. For example, 2-butene has the double bond in the center, and the terminal methyl groups can be on the same side (*cis*) or on opposite sides (*trans*) as in Figure 17.3. Again, these two isomers have very different properties, such as melting point and boiling point.

The more precise term to describe this type of relationship is *diastereomers*, stereoisomers that are not mirror images. Diastereomers tend to have very different physical properties, while enantiomers have identical physicals properties except for optical activity.

All right, let's now take a look at the results of a number of seemingly related reactions that will help us understand the state of organic reactions in the late 1950s and early 1960s. The reactions in Figure 17.4 involve the opening of a ring to produce an acylic compound with multiple double bonds. Although multiple products are in principle possible, these reactions net but a single product.

In the reaction shown in Figure 17.4a, this *cis* 3,4-dimethylcyclobutene can open to form a molecule with two double bonds. In principle, we might expect to observe a mixture of three different diastereomers, one with two *cis* double bonds, one with two *trans* double bonds, and one with a *cis* and a *trans* double bond. However, the only product observed has one *cis* and one *trans* double bond. Similarly, the ring opening reaction of the *trans* 3,4-dimethylcyclobutene (Figure 17.4c) produces predominantly the product with two *trans* double bonds and a very small amount of the isomer with two *cis* double bonds. Why are these reactions so specific?

It just gets more confusing when one considers the ring opening of the *cis* and *trans* cyclohexadienes, Figure 17.4b and 17.4d. In these examples, the products contain three double bonds, and the central one is always *cis*. But

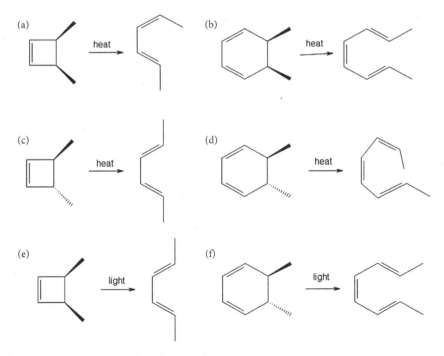

**Figure 17.4.** Some examples of pericyclic reactions.

for the other two double bonds, the *cis* reactant leads to the product with two *trans* double bonds, while the *trans* reactant leads to the product with one *cis* and one *trans* double bond, opposite the result with the cyclobutenes.

To muddy the waters further, consider that if these reactions are performed using light instead of heat to provide the activation energy, the products observed are the opposite. So, while heating *cis* 3,4-dimethylcyclobutene gives the product with one *cis* and one *trans* double bond (Figure 17.4a), shining light on this reactant produces the product with two *trans* double bonds (Figure 17.4e). Similarly, the reactions in Figure 17.4d and f produce different diastereomeric products, depending on whether heat or light is employed.

At first blush, we might fall back on the arrow-pushing notation to help us understand these reactions. So, for the reactions involving cyclobutenes, we might draw two arrows inside the ring, with one of the arrows indicating the breaking of the ring and forming a double bond, and the second arrow moving the double bond to an adjacent position (Figure 17.5). For the reactions of cyclohexadienes, we use three arrows showing the movement of the electrons around the ring, breaking the ring and making three double bonds.

In these four mechanisms, there is no source or sink of electrons; rather, electrons move around a circle. This arrow-pushing argument offers some hint that might guide us as to why these reactions produce different products—one involves four electrons moving around a circle and the other involves six electrons. But what this difference means was opaque until the 1960s!

These "arrows in a circle" mechanisms suggest that they are related to other well-known reactions. The reactions shown in the left side of Figure 17.6

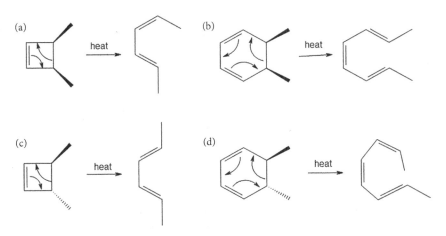

**Figure 17.5.** Arrow-pushing mechanisms involving four electrons (a and c) or six electrons (b or d).

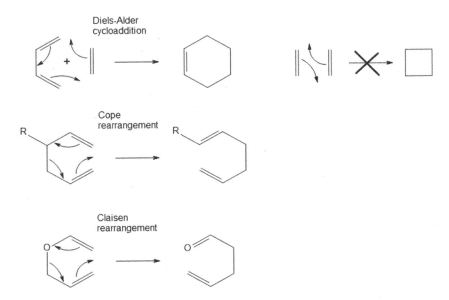

**Figure 17.6.** Examples of cycloaddition and sigmatropic rearrangement reactions.

represent examples of three very versatile reactions: the Diels-Alder cyclo-addition that takes two molecules and creates a six-membered ring, and the Cope and Claisen rearrangements that change the attachment points in alkenes. Each of them can be written with a single step where six electrons (three arrows) move around a circle. Particularly noteworthy is that the analogue of the Diels-Alder reaction whereby two alkenes combine by moving four electrons around in a circle to prepare a four-membered ring (right side of Figure 17.6) does not take place, except with the intervention of light.

The last two examples of this family of reactions are shown in Figure 17.7. Similar to the reactions in Figure 17.6, they can be written with a mechanism whereby six electrons move about a ring. These two examples represent what are respectively called *ene* reactions and *chelotropic* reactions.

All of these example reactions, and there are many, many thousands more, can be broadly classified as *pericyclic reactions*, reactions involving a single step with electrons moving around a circle. As discussed above, many pericyclic reactions proceed with stereoselectivity. The situation stood in the early 1960s with no organizing principles or predictive theory.

This confusing set of reactions plagued physical organic chemists for many years. (It was worse for organic chemistry students who had to resort to memorization of these reactions and outcomes.) These reactions look so similar,

ene reaction

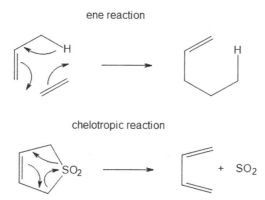

chelotropic reaction

**Figure 17.7.** Examples of ene and chelotropic reactions.

as do their outcomes but often with dramatic, repeatable differences. Many hypotheses were proposed, to no avail.

It was not until the mid-1960s, when the pioneering work of Robert Burns Woodward and Roald Hoffmann brought all these reactions under one consistent rationale: that of conservation of orbital symmetry. Around the same time, Kenichi Fukui used frontier molecular orbitals, and Howard Zimmerman applied ideas of aromaticity that contributed complementary frameworks to the Woodward-Hoffmann rules. These works revolutionized physical organic chemistry in important ways, but before considering the impact of these theories, I will first provide a taste of how these reactions were wrangled into submission. Full treatment requires serious application of quantum mechanics, which I will forego, settling for a more intuitive appreciation.

The example I will walk us through is for the ring opening of cyclobutenes, as shown in Figure 17,5a and 17.5c. The reverse reaction, the ring closure of butadienes, goes via the same mechanism, and so I will discuss both the forward and reverse reactions interchangeably.

In the cyclization of butadiene to cyclobutene, the two double bonds at the end break apart and a new double bond forms in the middle, while the end carbons bond together. Within quantum mechanics, bonds are described by *molecular orbitals*, the regions in space where the electrons are likely to be found. Molecular orbitals are composed of atomic orbitals, the s- and p-orbitals you may recall from an introductory chemistry class. Atomic orbitals describe the space where electrons are likely to be found within an atom.

A single bond is described by a molecular orbital formed of a p-like orbital on each atom pointing at each other, as in Figure 17.8. The p-like atomic

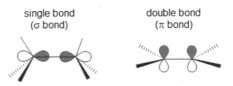

single bond
(σ bond)

double bond
(π bond)

**Figure 17.8.** Orbital picture of the σ and π bonds.

conrotatory

disrotatory

**Figure 17.9.** Orbital model for the creation of a σ bond through a conrotatory or disrotatory path.

orbital looks like a dumbbell with each side having a different mathematical sign; one side is positive and one side is negative. This is pictorially represented with one lobe colored in and the other empty. Note that the bond is formed by the in-phase overlap of the atomic orbitals, meaning that the same signs (i.e., the same color) of the lobes are pointing at each other. This head-to-head arrangement of the atomic orbitals forming a single bond is also referred to as a σ-bond.

The double bond is made up of one σ-bond and one π-bond. The π-bond is described by a molecular orbital formed of a p-orbital on each neighboring atom arranged parallel and in phase, as in Figure 17.8. The lobes on the top have the same sign as do the lobes on the bottom.

So from a molecular orbital perspective, the cyclization of butadiene to cyclobutene proceeds by the end carbons rotating so that the p-orbitals that are oriented up and down end up pointing at each other to form the new σ-bond that closes the four-membered ring. This rotation can take place in two ways. First, the two carbons can both rotate in the same direction, say, both can rotate clockwise, as shown in Figure 17.9. This is called a *conrotatory* process. The alternative is for one carbon to rotate clockwise and the other to rotate counterclockwise, also shown in Figure 17.9. This is the *disrotatory* process. Both allow for the cleavage of the two terminal π-bonds and the creation of the σ-bond.

What may not seem obvious at first glance is that we can readily distinguish the result of the conrotatory from disrotatory path. Suppose we label the two positions that are in the interior region of butadiene and the two positions that are on the outside. I have done this symbolically in Figure 17.10 using "I"

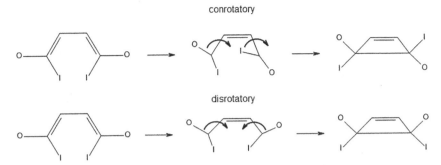

**Figure 17.10.** Tracking the differences in a conrotatory and disrotatory path for ring closure of butadiene.

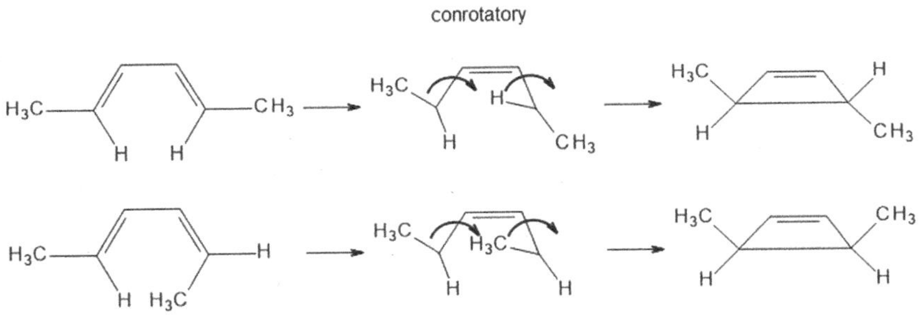

**Figure 17.11.** Conrotatory ring closure of butadienes to cyclobetenes.

and "O." but synthetic chemists have many techniques for attaching specific atoms or groups at these positions to create real molecular examples.

Let's follow the outside groups. In the conrotatory pathway, the outside group of the left terminal carbon heads to the top, while the outside group on the right terminal carbon heads toward the bottom, resulting in the product having the groups on opposite sides, or the *trans* product. For the disrotatory pathway, both outside groups move toward the top, leading to the *cis* product. These two stereoisomers are readily distinguishable by experiment.

So, what is the actual outcome of this type of ring closure (also called *electrocyclization*)? If we examine Figure 17.5c, butadiene with two methyl groups in the outside position end up in a *trans* arrangement in the cyclobutadiene product. That corresponds to the conrotatory pathway in Figure 17.11. Similarly, the example in Figure 17.5a takes the butadiene with one group inside and one group outside into the *cis* product. That too corresponds with the conrotatory pathway (see Figure 17.11).

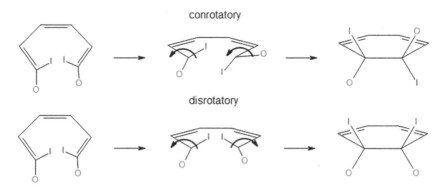

**Figure 17.12.** Tracking the differences in a conrotatory and disrotatory path for ring closure of hexadiene.

The same analysis can be carried out for the electrocylization of a hexatriene to a cyclohexadiene (see Figure 17.5b and d). By designating the inside and outside positions, both conrotatory and disrotatory paths can be discerned (Figure 17.12). Close inspection of the known electrocylization reactions, such as those shown in Figure 17.5, reveals that the disrotatory pathway is the one that takes place.

Looking back on Figure 17.5, the two electrocyclization reactions involving two curved arrows (four electrons) follow the conrotatory pathway. On the other hand, the electrocyclization reactions that involve three curved arrows (six electrons) follow the disrotatory pathway. Is that just coincidence? Or is there some deeper explanation than can rationalize these results?

The answer lies with the nature of the molecular orbitals of the reactant and product. You may have perhaps noticed that the molecular orbitals displayed in Figure 17.9, which demonstrate the formation of the σ-bond, are different for the conrotatory and disrotatory pathways. That's because the symmetry of the orbitals must be different in order to allow for the in-phase combination of the p-orbitals. For the disrotatory path, the tops of the p-orbitals must be in phase, while for the conrotatory path, the lobes from opposite sides must be in phase. Woodward and Hoffmann built on this notion of symmetry to flesh out a full set of rules for these types of reactions.

Next, I present a sense of their arguments for how molecular orbital theory explains pericyclic reactions. Figure 17.13 presents the most important molecular orbitals for butadiene and cyclobutene, the reactant and product of the electrocyclization reaction. These are the so-called *frontier molecular orbitals*, the orbitals involved in the bonding changes in the reaction. These include the highest energy orbitals that are occupied by electrons.

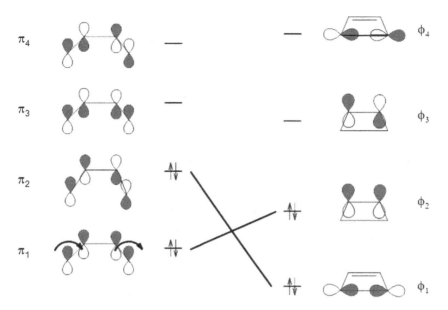

**Figure 17.13.** Molecular orbital diagram for the conrotatory electrocyclization of butadiene into cylobutene.

Low-energy unoccupied molecular orbitals are also included in the frontier orbital collection. Unoccupied orbitals are important as they provide the space where electrons can move. This can be seen in two important reactions. First, electrons can be added to a molecule. The occupied orbitals contain all the electrons they can house, which is a maximum of two per orbital. The lowest energy unoccupied orbital is then the optimal place for an added electron to go. Second, energy can be added to a molecule, such by exposing it to light. The energy in light can be *absorbed* by exciting an electron from a low-energy occupied orbital to move into a higher-energy unoccupied orbital. This creates an *excited state* molecule, a molecule more energetic than usual. Both of these processes are widely used within chemistry.

In Figure 17.13, I have displayed the two highest energy occupied and the two lowest energy unoccupied orbitals for butadiene and cyclobutene. Next, consider how the orbitals change during the conrotatory process. When the terminal carbon atoms rotate in a clockwise manner, the p-orbitals in the lowest energy orbital of butadiene (labeled $\pi_1$ in Figure 17.13) rotate so that the lobes on opposite sides overlap; and this is in an out-of-phase way, making no C–C bond. The two p-orbitals of the central two carbon atoms remain in phase, making a $\pi$-bond. This means that during the conrotatory process the butadiene $\pi_1$ orbital will change into the $\phi_2$ orbital of cyclobutene. When the two terminal p-orbitals of $\pi_2$ rotate clockwise, the lobes on the opposite side now overlap in an in-phase way, making the new C–C bond. So, the $\pi_2$ orbital changes into the $\phi_1$ orbital. Overall, in the conrotatory process the two lowest energy orbitals of the reactant change into, or *correlate* with, the two lowest energy orbitals of the product.

What about for the disrotatory pathway? The exact same orbitals for cyclobutene and butadiene are reproduced in Figure 17.14, but now we consider how they transform when the terminal carbons rotate in opposite directions. The two terminal p-orbitals in $\pi_1$ will rotate so that the top lobes come together in phase, making the C–C bond, which is the $\phi_1$ orbital of cyclobutene. But when the p-orbitals of $\pi_2$ rotate in opposite directions, their top lobes overlap out of phase, and the p-orbitals of the middle two carbons are also out of phase. That correlates to the cyclobutene $\phi_3$ orbital. This means that the electrons that start out in $\pi_2$ would end up in a higher-energy orbital ($\phi_3$) than ideal ($\phi_2$). The disrotatory pathway takes butadiene into an excited state of cyclobutene, not its ground, or most stable, state.

It's this subtle difference in orbital correlation that provides the explanation for the selectivity in this reaction. The conrotatory process allows the ground-state reactant to end up in the ground-state product. The disrotatory process takes the ground-state reactant into the excited state product, which would

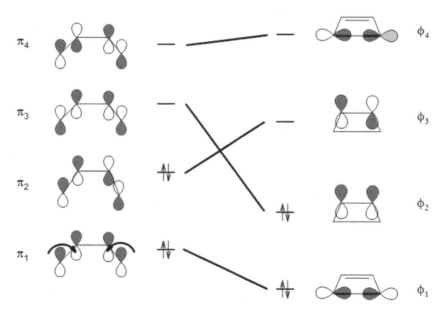

**Figure 17.14.** Molecular orbital diagram for the disrotatory electrocyclization of butadiene into cyclobutene.

be higher in energy, and less stable. Woodward and Hoffmann called the conrotatory process *allowed* and the disrotatory process *forbidden*. This latter term is a bit unfortunate in that "forbidden" implies "never," while in actuality, all that quantum mechanics can really say is that the disrotatory pathway here is very unlikely to happen. This explanation is consistent with numerous experiments.

A similar molecular orbital analysis of the reaction of substituted hexatriene to substituted cyclohexadiene (as in Figures 17.5b and d) finds that the allowed reaction is for the disrotatory pathway and that the conrotatory pathway is forbidden. This result is also consistent with many experiments.

Woodward and Hoffmann applied this type of analysis to a wide variety of pericyclic reactions, including the ones shown in Figures 17.6 and 17.7. They were able to summarize their results into some simple rules. For example, pericyclic reactions involving the movement of $4n$ electrons (those whose mechanisms have two curved arrows, or four curved arrows, etc.) will follow a conrotatory pathway. Those pericyclic reactions involving the movement of $4n + 2$ electrons (those whose mechanisms have one curved arrow, or three curved arrows, etc.) will follow a disrotatory pathway.

Returning to Figure 17.4, we see that reactions carried out with light (*photolysis*) instead of heat (thermolysis) proceed with opposite stereochemistry. So, for the reactions of the cyclobutenes (as in Figure 17.4e), photolysis proceeds through a disrotatory pathway, opposite to what happens in thermolysis; and the same switching between photolysis and thermolysis happens with the cyclohexadienes (as in Figure 17.4f).

Photolysis causes an excitation that moves one or more electrons into a higher lying orbital, usually into the lowest unoccupied molecular orbital. If we then consider how this orbital correlates under conrotatory or disrotatory motion, we see that the opposite selection rules apply, namely, that $4n$ electron reactions will proceed via the disrotatory path and $4n + 2$ electron reactions will proceed via the conrotatory path. Similar orbital analysis finds that the Diels-Alder reaction is allowed, but the reaction of two double bonds to form a four-membered ring (see the right side of Figure 17.6) is forbidden. With light, this forbidden reaction becomes allowed! Numerous examples of this photolysis reaction have been reported.

Fukui's frontier molecular orbital theory complements the work of Woodward and Hoffmann. Frontier molecular orbital (FMO) theory is particularly adept at explaining and predicting the regioselectivity of some pericyclic reactions like the one shown in Figure 17.15. Note that the two molecules can come together in two relative orientations: where the two substituents end up opposite to each (70%) and where the substituents are separated by just

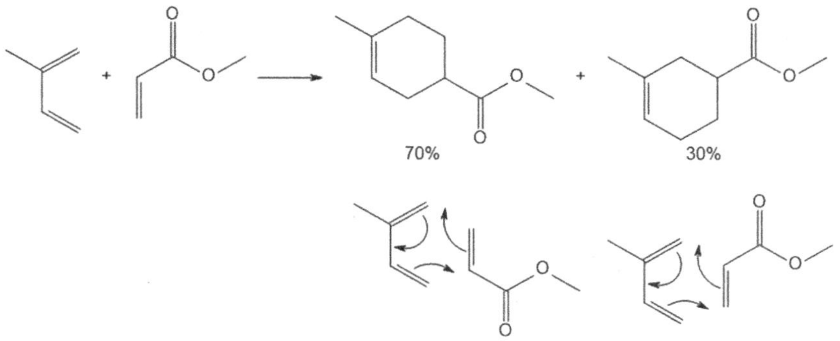

**Figure 17.15.** Regioselectivity in a Diels-Alder reaction.

one carbon (30%). FMO arguments are based on matching up the orbitals on each molecule such that the biggest lobes interact, and that can explain the outcome of the reaction shown in Figure 17.15 and many other ones as well.

Zimmerman's approach divides pericyclic reactions into those that follow a Hückel arrangement, where the reacting orbitals all line up parallel, and a Möbius arrangement, where the orbitals have a single twist. Fundamentally, all three approaches rely on similar grounding in quantum mechanics, and the three approaches make the same predictions. Hoffman and Fukui shared the 1981 Nobel Prize in Chemistry for this work; Woodward would undoubtedly have shared in the Prize as well had he not passed away a few years earlier.

The power of the Woodward-Hoffmann rules is its scope. These rules apply to all manner of pericyclic reactions, hundreds of thousands of them! All of the types of pericyclic reactions shown in Figures 17.5–17.7 are brought together within this theory. That's one of the attributes of a paradigm-shifting theory: a theory that brings together a broad swath of experiments, often seemingly unrelated to each other, that can now be understood within a simple framework or rationale.

Woodward and Hoffmann were incredibly brash about the scope of their theory. In their important book summarizing their work, *The Conservation of Orbital Symmetry*, Chapter 12, which is entitled "Violations," begins with the quote:

> *There are none! Nor can violations be expected of so fundamental a principle of maximum bonding.*

While much has been made of this statement, and many attempts have identified seeming "violations," in fact the theory is well grounded. But that does not mean it is as absolute as newcomers might think. What the symmetry rules identify are relative activation barriers, not what reactions can take place and what reactions cannot take place. Certainly, when we have situations of differing barrier heights, supplying more energy can get all reactions to occur, as we shift from kinetic to thermodynamic control where other outcomes may prevail. Quantum mechanics rarely forbids anything from happening; rather, being probabilistic, quantum mechanics suggests that some outcomes are most unlikely. That positions the work of Woodward and Hoffmann as telling us what outcomes are more likely than others—and that information has been exploited for decades now by organic chemists in their designs for the synthesis of complex molecules.

Perhaps the more important outcome of the orbital symmetry theory lies in the cultural change it brought about. The great success of the Woodward-Hoffmann rules established the central role of the mechanism as a way of understanding the "how and why" of a reaction. Take, for example, the two cyclization reactions shown in Figure 17.16. The top reaction is an example of a rather ordinary Diels-Alder reaction, involving six electrons. The bottom reaction is a bit more esoteric, involving sixteen electrons moving around a circle. The Woodward-Hoffmann rules explain the different

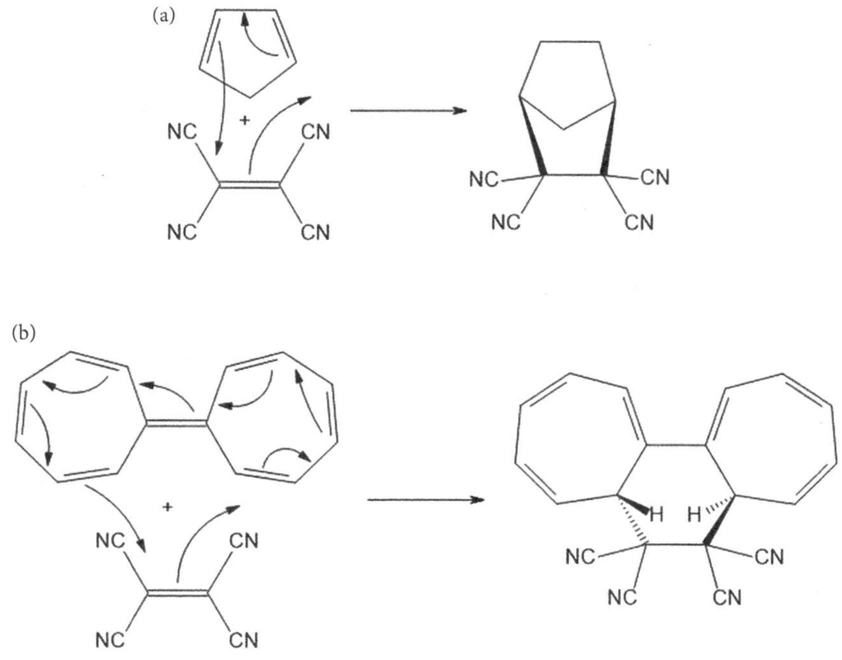

Figure 17.16. Cycloaddition reactions involve 6 and 16 π electrons.

stereochemistry: the top reaction, having $4n + 2$, electrons, adds the molecule *cis*, while in the bottom case, involving $4n$ electrons, the resulting stereochemistry is *trans*.

For that that second reaction, more compelling are all of those arrows moving those many electrons to create the new bonds. All those predicted new bonds match up precisely with the experiment result. The ubiquity of the success of the arrow-pushing model for pericyclic reactions spurred all organic chemists to consider arrow-pushing arguments for all reactions.

In preceding chapters, I've shown examples of the utility of arrow pushing in rationalizing substitution and elimination reactions. A couple of examples of more complicated mechanisms, explained with a series of arrow-pushing steps, are shown in Figure 17.17. Both involve multiple steps, each understood with bond changes depicted by the arrows moving from some source of electrons to some sink. These examples are by no means the most complicated mechanisms that physical organic chemists have developed. All the same, they demonstrate the level of complexity and creativity that nature (and physical organic chemists) can explore.

The power of the Woodward-Hoffmann rules to bring together such a large group of reactions under one theory was overwhelming. If we can create these rules, which are frequently put into place by depicting the arrow pushing, why not do this for everything else too?

In the 1960s and 1970s, physical organic chemistry hit its heyday, with many research groups scattered across the globe providing experimental data to validate the Woodward-Hoffmann rules and to extend the notion of arrow-pushing mechanisms to more and more reactions. Conference after conference was dominated by physical organic chemists detailing their latest discoveries that provided ever-growing insight as to how reactions were proceeding.

Eventually, this created a sense of organic chemical intuition. Organic chemists will recognize reactive centers: atoms that are prone to attack by an electrophile (typically a negatively charged atom or an atom with a lone pair of electrons), or atoms that are prone to attack by a nucleophile (typically, a positively charged atom). We identify groups of atoms that behave similarly, regardless of what else is attached to the molecule. We note arrangements of multiple bonds and the patterns of reactions these multiple bonds engage in. We take note of the pH and allow protons to jump in when we have a reaction in acidic medium and hydroxide ions (HO⁻) to participate in reactions carried out in base. We look for opportunities to create aromatic rings; we avoid constructing strained rings, and we will bust apart a small ring to help drive a reaction forward.

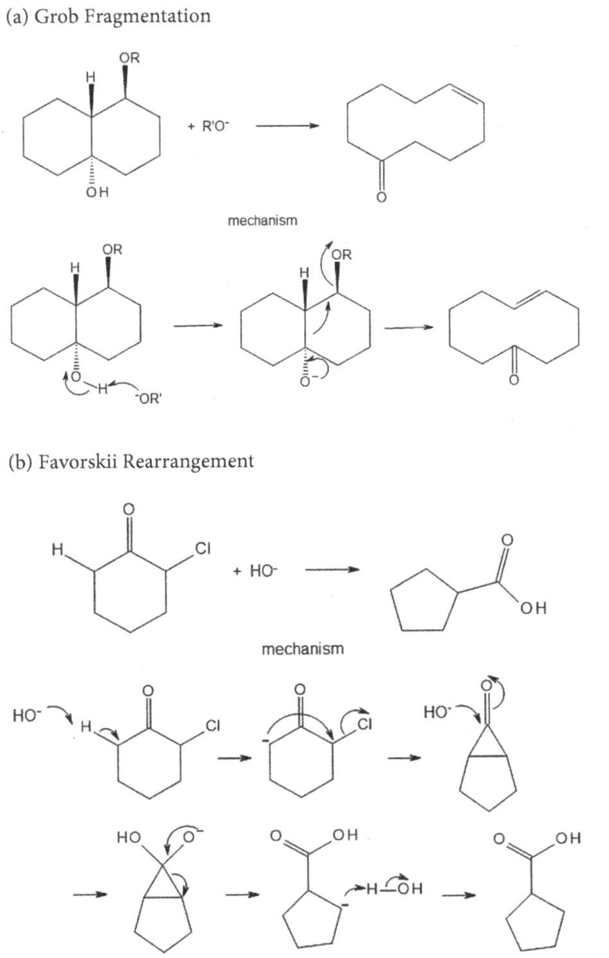

(a) Grob Fragmentation

(b) Favorskii Rearrangement

**Figure 17.17.** Mechanism of the (a) Grob fragmentation and (b) Favorskii rearrangement.

After Woodward and Hoffmann's publications, arrow pushing became *de rigueur* when any new organic reaction was published. Reporting a new synthetic reaction was accompanied by a proposed mechanism. It was as if an experimental procedure could be true only if we had an accompanying mechanism to rationalize it.

For many decades, the initial report of a new mechanism would likely include some experimental data to support the proposed (arrow-pushing) mechanism. But over time, the proposed mechanism was often supported simply by chemical intuition. Some of this change was due to the success of physical organic chemists to construct this edifice of arrow pushing. Another cause of this change was the dwindling supply of grant funds available for traditional physical organic chemistry.

But the cultural change within the organic chemistry community was profound. Organic synthesis transformed from a discipline that was largely art and happenstance to one of prediction and execution. Previously, reactions were developed through analogy and trial-and-error. Predictability was limited and based on empiricism, an appeal to prior examples. Subsequent to the publication of the Woodward-Hoffmann rules, organic synthesis often became reliably predictable. Arrow pushing provided a means for understanding reactions and offered insights as to how to alter conditions to change outcomes. Mechanism became the *lingua franca* of organic chemists.

The reach of the arrow-pushing and reaction mechanism has grown outside organic chemistry. Biochemical reactions, reactions that take place inside a cell, are often well-understood organic reactions that just happen to take place within large molecules. But some processes are more complicated, and enzyme activities and metabolic processes are now described using the language of organic mechanisms: arrow pushing. Similarly, inorganic chemical reactions, reactions that generally do not involve reactions at carbon atoms, are now also discussed using the arrow-pushing paradigm.

Becoming adept at arrow pushing is a key element to success in an introductory organic chemistry class. It is the way to organize the myriad reactions and bring some semblance of order to the material. Once a student becomes comfortable with this process, new reactions can be incorporated into one's arsenal of reactions in a seamless way; it's not really all new material anymore, it's simply an extension made using an already known framework. And we have the work of Woodward and Hoffmann on pericyclic reactions to thank for this!

# 18
# Reaction Dynamics

You are driving down a road and have to make a right turn. As you enter the turn, you realize that it is tighter than you anticipated. You gently hit the brakes, turn the wheel more to the right, and continue safely onward.

Now imagine this same scenario on a winter's day. The road is covered with snow and ice. As you prepare to begin the turn, your car starts to skid. It continues to go mostly straight ahead, but it refuses to go to the right despite your efforts in turning the steering wheel. Tapping the brakes makes no effect. It's as if your car has a mind of its own, moving uncontrollably forward. The car continues skidding straight ahead until it runs up over the curb and finally comes to a stop after hitting a lamppost.

During artillery training, the new recruits practice use of the mortar. The team adjusts the angle of the tube and its orientation to try to deliver the mortar onto the target. Once they have made their adjustments, the mortar is dropped down the tube and launched skyward. It follows a graceful arc, but on its way downward, the team realizes that the mortar will fall short of the target. However, there is nothing they can do to affect the path of the projectile. Its path was set at launch and cannot be altered.

Recall the scene near the end of the journey made by Tom Hanks' character aboard the raft in the film *Cast Away*. The raft is just a few logs at this point. Wilson, his volleyball companion, is long gone, and Hanks' character, Chuck Noland, is dehydrated and awaiting his final fate. And then we hear the horn and see a giant container ship move by him in the background. What we don't see is what happens next in order to save him. This large ocean vessel cannot stop on a dime. The container ship will need miles to stop in order to launch a small lifeboat to return to rescue him. Chuck's salvation will take hours.

What do these stories have in common? Momentum—the physical property of an object that describes a combination of its mass and its velocity. Isaac Newton told us that an object in motion tends to stay in that motion until it is disturbed by some force. This tendency is the particle's momentum. The greater the momentum, the more difficult it is to move the object onto some new path.

*Thinking Like a Physical Organic Chemist.* Steven M. Bachrach, Oxford University Press. © Oxford University Press 2023.
DOI: 10.1093/oso/9780197640371.003.0018

Since momentum is defined as the product of an object's mass and velocity, any increase in either component results in increased momentum, and thereby makes the object more resistant to change in its motion. So, a fast-moving object will have a larger momentum than a slower moving object. That's why applying the brakes and slowing the car allow you to navigate the turn in the first scenario. The slower car has less momentum, making it easier to alter the path and turn the car to the right. A heavy object has a large momentum. That's why the container ship will take a long time and travel quite a distance before coming to rest.

In the absence of some force, an object's momentum will keep it moving along its set path. Once the mortar is fired, the only force acting on it is gravity, which changes the path into an arc to fall back to earth. There is no way to redirect the mortar, no way to speed it up or change its motion to any side. The car skidding on ice acts that way because the ice dramatically decreases the friction with the ground, reducing the force that allows the car to turn or stop.

In classical mechanics, knowledge of the potential energy surface, the position of every object, and the momentum of every object is needed to make useful predictions. With this information in hand, one can use Newton's law of motion ($F = ma$) to trace out the time evolution of any system of objects. For example, if you know the positions of all of the balls on a billiard table, the velocity and mass of all of the balls, and the force of the friction of a ball interacting with the felt on the table, you can predict the travels of each ball. You can compute every collision between the balls, the bounce of each ball off the rails, and which balls will fall into a pocket.

The path an object takes is called a *trajectory*. *Dynamics* is the computational process of tracing out the trajectory of an object or objects under study, whether this be the balls on a billiard table or the planets in the sky. *Molecular dynamics* is the exact same computational process, only applied to the motion of atoms and molecules.

If you have watched any movie about NASA, such as *Apollo 13* or *Hidden Figures*, you have undoubtedly seen a plot of the trajectory of the rocket from launch travel around earth or to the moon and back. That plot, that trajectory, was computed using dynamics through application of Newton's laws. Entered into the computation were the location of the launch pad, the weight of the rocket, the thrust of the engines, the velocity of the earth, moon, and sun, the force of gravity due to these bodies, the effects of general relativity, and the drag due to the atmosphere. Effectively, the initial condition of the rocket, the earth, moon and sun, and the forces that will act upon all of these objects are put into the computer, and then the positions of all of these objects are computed for some small amount of time into the future. This procedure is then repeated over and over again, plotting out the motion over time of the rocket, the earth, the moon, and the sun. With a correct computation, the astronauts land at the right place on the moon and return to the right place on earth.

These calculations find the trajectory in phase space of every object on a given potential energy surface. *Phase space* accounts for the position and momentum of each object at every point in time.

In all of the discussions of organic reactions so far, I have considered the position of the atoms in a molecular system moving upon a potential energy surface. I presented reactants whose atoms move along the reactant coordinate in such a way as to climb up the barrier and across the transition state, and then move downhill to an intermediate with some new bonds made and some old bonds broken. Conspicuously absent was any discussion of the momenta of the atoms and the time evolution of the positions and momenta.

In a sense, we have ignored fully half of the physics! Clearly, momentum matters in our everyday lives, as exemplified by the scenarios at the start of this chapter. By not considering momentum, we tacitly assumed that the path the atoms and molecules actually take is irrelevant! Shouldn't paths matter for atoms and molecules too?

For many situations paths do not matter, even in our ordinary lives. We really don't care how our gift package travels from our home to our parents' house across the country. The package can go by air or rail or truck, just as long as it arrives intact and on time. The actual path the package traversed is unimportant. When we are out for dinner with friends, the order in which the kitchen prepares the different dishes does not matter; we only care that all meals arrive well prepared and served together. When completing our taxes for the Internal Revenue Service (IRS), we combine all wages together on one line, regardless of which partner earned wages or how many jobs one had during the year.

On the other hand, for many situations the path is of critical importance. The CFO of FedEx certainly worries about how to cost-effectively deliver our package across the country. The mode of transportation and the route directly figure into the bottom line of the company. The head chef at the restaurant carefully choreographs when each dish is cooked, so that the meals are delivered to the table all at once, insuring that the diners are thrilled with the meal and will return often. The IRS does not consider all income equivalently; for example, it taxes wages at a different rate than it does capital gains. In other words, the way an individual accrues the total income affects the amount of taxes paid.

In chemistry, too, some measurements are path-independent and some are highly sensitive to the path. If we are interested in how much energy is released when a reaction takes place, like when we burn gasoline in a car engine, the path does not matter. In contrast, the rate of a reaction is dependent on the path. A large activation barrier results in a slow reaction. The use of a catalyst will change the reaction path, providing an alternative path with a barrier that is much lower in energy. An older car will produce harmful pollutants such as nitrogen oxide (NO) and carbon monoxide (CO) as part of the combustion process. These gases would just flow out of the tailpipe and into the environment. Both of these gases can be converted to nontoxic species like nitrogen gas ($N_2$) and carbon dioxide ($CO_2$), but these reactions are exceedingly slow, measured in years. The catalytic converter on modern cars contains platinum, palladium, and other metals that allow rapid conversion of the harmful gases into benign ones, such that almost no NO or CO is emitted from the tailpipe. By providing an alternative reaction pathway, these environmentally critical reactions are accelerated by the metal catalysts.

For some situations, chemists can ignore specific pathways by treating the system with statistics. I have mentioned this idea once before as the way that Maxwell and Boltzmann developed their theories to explain the behavior of gases. Instead of trying to track the path of every molecule in a gas, they applied a statistical treatment that allows for significant simplifications in the computations.

Statistics works best if the sample size is very large and the distribution is known or predictable. For the gas problems, the sample size is enormous. If you recall the *mole* from an introductory chemistry class, you'll remember Avogadro's Number as $6.02 \times 10^{23}$, and that's a typical sample size. For a regular distribution, we need to consider the time scale of vibrations and rotations of molecules, and these are typically on the order of a nano- to microsecond. In most laboratory experiments, the molecules have ample time to achieve a stable, regular distribution.

Physical chemists have developed two powerful theories for understanding and predicting reactions rates: *transition state theory* (TST) and *Rice–Ramsperger–Kassel–Marcus theory* (RRKM). Both theories assume statistical treatment of all the chemical species involved in the reaction, including any intermediates. That means that, for example, an intermediate exists for long enough that all of its vibrations and rotations can settle into a normal (Boltzmann) distribution. Since an intermediate might be formed in such a way that excess energy is in one or two vibrational modes, it will take some time, say, a few microseconds, for that excess energy to redistribute into the other vibrational modes.

If this redistribution can occur, then these statistical theories work extremely well. And that's what allows us to often neglect the actual reaction pathways and momenta. The statistical treatment essentially averages over all of the paths, and we need not consider any specific path at all. Until about twenty years ago, most organic chemists comfortably made this assumption and happily discussed reactions and wrote textbooks focusing on potential energy surfaces and atomic positions, while neglecting momenta entirely.

Let me provide a simple set of *gedanken-experiments* that suggest why molecules might behave in a nonstatistical way. Using Figure 18.1 as a guide, imagine a skier at the top of a hill (P1), looking down on a slope that first enters a deep gully (P2), crosses a small hill (P3), and ends on a flat plain (P4 and P5). Note that the starting hill is much higher than any other of our identified points and that the ending flat plain is a bit higher than the gully. Our skier is capable of only providing an initial push off the mountain, and from then on, the skier is in a tuck and does not use the poles or any other means to change speed or direction.

Let's walk through three different scenarios our faithful skier might experience. The first scenario has the surface covered with highly polished ice, and the air is very still. Our skier pushes off, heads down the first slope, picking up speed all the way down, shoots up the second hill, and then down onto the final plain, continuing on for quite some distance, coming to rest at P5.

In the second scenario, the course is covered with a nice powdery snow. The skier comes down the first hill and enters the gully with less speed than in the first scenario but can easily go up and over the hill (P3) and comes to a stop at P4.

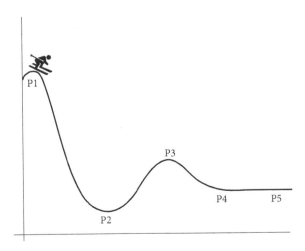

**Figure 18.1.** Ski slope analogy.

In the last scenario, the surface is covered with poor snow, with some grass and rocks poking through. Our hapless skier creeps down the first hill never really gaining much speed, enters the gully, and cannot climb up the second hill, ending at P2.

To understand these three scenarios, we need to apply the law of conservation of energy. When the skier is at the top of the hill at P1, she's at rest and has no kinetic energy, but lots of potential energy is tied up in gravity. As she drops down the hill, the loss of potential energy (gravity) must be converted into some other form of energy. One way to conserve energy is to convert it into kinetic energy and have the skier go faster and faster as she drops down the hill. But the skier is also interacting with the surface of the slope and can lose energy to friction. As the skier progresses down the slope the potential energy goes into making the skier move faster, and when she goes uphill, she loses speed to convert the kinetic energy into potential energy. And all the while there is steady loss of energy to friction.

For the first scenario, the polished ice surface means that we can just about neglect friction. For the most part, we need only consider the interchange of potential and kinetic energy. As the skier drops down the slope, her speed will increase as the potential energy is converted into kinetic energy. When she starts to climb up the P3 hill, her kinetic energy is converted into potential energy, but she has so much excess kinetic energy that she can easily go over the hill and continue on the plain. She will come to a rest only because the slow loss of energy to friction will eventually bring her to a stop.

In the second scenario, the powder snow will cause real loss of energy to friction throughout her trip. Nonetheless, this loss of energy is slow enough that she has sufficient speed, and that means sufficient kinetic energy, to still cross the P3 hill and come to rest at P4.

The very poor snow cover in the last scenario results in rapid loss of kinetic energy to friction. As the skier moves down the slope, she gains some kinetic energy, but that quickly dissipates as friction. She moves very slowly down the hill, and when she reaches the bottom (P2), she is moving quite slowly, having only a small amount of kinetic energy. She lacks the energy to climb up over the P3 hill, and so she stops in the gulley.

Let's now use these scenarios to consider what happens with molecules. I will now redraw the skier surface for molecules as shown in Figure 18.2. Importantly, I have labeled the $y$-axis as the sum of the potential and kinetic energy. In all of the previous plots, the $y$-axis has been just the potential energy. By including the kinetic energy, this plot has subtly changed to consider momentum as well as position. Let's also consider the molecule starting at the

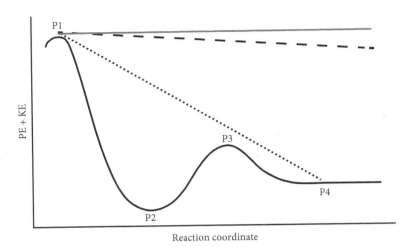

**Figure 18.2.** Molecular analogue of the ski slope in Figure 18.1. The gray line is the trajectory with no loss of energy to friction. The dashed line is with slow loss of energy as friction, that is, few collisions. The dotted line is with moderate loss of energy as friction, that is, moderate collisions.

position of the first transition state at the left, with just a bit of kinetic energy to push it forward.

Conservation of energy suggests that as the molecule moves forward along the reaction coordinate, the sum of the kinetic energy and potential energy will remain constant; this is displayed as the top, horizontal gray line in Figure 18.2.

But our three skier scenarios had one other type of energy to consider: friction. Is there a molecular analogue of friction, some way for kinetic energy to be removed from the molecule? The answer is yes. The reacting molecule can bump into neighboring molecules and transfer some energy through each collision. Collisions happen within gases, and collisions are even more common in solution where the molecules are closer to each other.

Our first scenario of the skier on a very icy slope is analogous to a reaction in a very sparse gas. Here collisions will be rare, and kinetic energy will only be slowly transferred to the gas through the rare collisions (the dashed line of Figure 18.2). The second scenario is more realistic, with more frequent collisions bleeding off kinetic energy. This is represented as the dotted line in Figure 18.2. The molecule ends up at P4 and simply passes over the region of the intermediate P2.

The third scenario is akin to the reacting molecule losing kinetic energy as quickly as it is gained. The molecule would follow the black line, the actual potential energy surface, since it never really possesses any significant kinetic energy—it has all been lost to friction via collisions. The molecule would end up at P2, as it would have no kinetic energy available to climb the second transition state.

Now this perfect loss of kinetic energy through collisions can never happen. There is just no way to perfectly couple the reacting molecule with the solvent molecules and bleed off the kinetic energy as rapidly as it is created.

It is more likely that some slow but steady loss of kinetic energy occurs as the molecules randomly bump into each other. But there seems to be a problem here connecting these situations with what we have discussed many times before. How does an intermediate exist with enough lifetime for the energy in all of the vibrations and rotations to become evenly distributed? The dotted line implies that the molecule just passes right through the intermediate with no lifetime. The black line (perfect loss of kinetic energy) implies that the intermediate is formed, but there is no way for it to pass over the second barrier; the intermediate would persist forever.

The typical situation is the following. The molecule crosses the transition state, and as it moves forward down the slope, potential energy is transferred to kinetic energy. The molecule picks up speed, but some of the kinetic energy manifests as increased vibrations of the bonds and rotation of the molecule. Additionally, some of this kinetic energy bleeds off in collisions with neighboring molecules.

As this vibrating, rotating, translating molecule enters the region of the intermediate, it likely has sufficient overall kinetic energy to surmount the second transition state. However, the molecule will need to have sufficient kinetic energy in the vibrational mode that will carry it up and over the transition state. It's analogous to a ball rolled forward up a hill. If it's thrown directly in line with the pass, it will climb up and over, following the path of the black line in Figure 18.3. But if the angle is off, the ball will roll up the hill, missing the pass, and come back to the same side, like the dashed path in Figure 18.3. In both throws, the ball has sufficient kinetic energy to surmount the pass. However, that's not sufficient for crossing over. The ball must be moving in the correct direction too. It's momentum that matters.

While a molecule can be in the intermediate region and have sufficient energy to cross over the activation barrier, it still needs to be moving in that proper direction to go over the hill. That proper direction is associated with a vibrational mode, and if there isn't enough energy in that specific vibrational mode, the molecule will bounce off the hill and remain in the intermediate well. It will bounce around in the intermediate well until its momentum lines up with the exit channel over the activation barrier. For the vast majority of

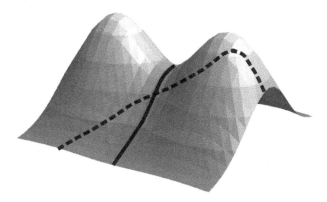

**Figure 18.3.** Two trajectories on a surface with two hills. The solid path follows the steepest ascent up through the pass, while the dashed path misses the pass and does not traverse to the other side.

diene          dienophile

**Figure 18.4.** Diels-Alder reaction of cyclopentadiene (diene) with ethylene (dienophile).

reactions, this lifetime in the intermediate well is long enough for statistical distribution of energy among the vibrational modes to occur, and then our statistical models will apply.

Research over the past twenty years has uncovered more and more examples of reactions that do not follow these statistical theories. Some reactions have rates much faster than expected. Some reactions result in an unexpected product. Some potential energy surfaces have features that were unexpected, and these can result in a different product than anticipated. I'll next present a few examples, and then I'll consider the broader implication of these studies, which have upended our traditional understanding of organic reactions.

As discussed in the previous chapter, the Diels-Alder reaction is the classic example of a pericyclic reaction. The reaction involves one reactant that contains a *diene*: two double bonds separated by a single bond. The second reactant contains a double bond (or sometimes a triple bond) and is called the *dienophile*. Figure 18.4 shows a Diels-Alder reaction of cyclopentadiene as the diene component with ethylene as the dienophile.

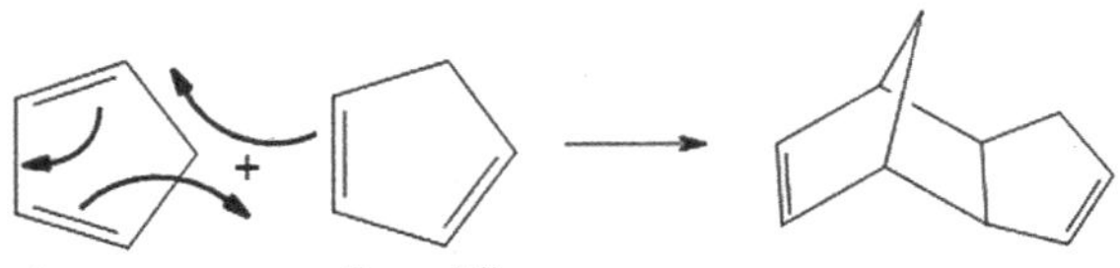

diene                    dienophile

**Figure 18.5.** Diels-Alder reaction of cyclopentadiene (diene) with cyclopentadiene (dienophile).

What if the dienophile also happens to have a diene component? For example, what if two cyclopentadiene molecule undergo a Diels-Alder reaction, as shown in Figure 18.5? This reaction is quite readily accomplished. The key interesting twist is that, unlike in the first example where ethylene can only act as the dienophile, this time each molecule can serve as either the diene or the dienophile.

Imagine if we could somehow track each specific molecule and thereby know which acted as the diene and which as the dienophile. Well, in principle we can do this through isotopic labeling. Suppose in one molecule, all of the carbon atoms were $^{12}C$, and in the other molecule, all of the carbons were the isomer $^{13}C$. The result would be two different products, one where the unlabeled cyclopentadiene is the diene component and the labeled cyclopentadiene is the dienophile, leading to product **A**, and the other where the roles are swapped, leading to product **B** (see Figure 18.6).

Of course, nothing is ever simple. One other reaction can occur, and that is another type of pericyclic reaction, the so-called Cope rearrangement. The two products **A** and **B** can interconvert through a Cope rearrangement, as shown in Figure 18.6c.

So what might we expect for the overall mechanism here? Well, our chemical experience would suggest that the reactants progress through either **DA-TS1** to give **A** or pass through **DA-TS2** to give **B**. And then **A** and **B** can interconvert by a Cope rearrangement through **Cope-TS**. This mechanism is depicted in Figure 18.7a.

Then in 2002, a group of Italian computational chemists led by Pierluigi Caramella applied quantum mechanics to this reaction and discovered that there is only one transition state for this Diels-Alder reaction. This result seemingly violates one of the principles for devising reaction mechanisms: that a transition state connects one reactant to one product! Instead, what we have here is a reactant connected to *two different products by one transition state.*

While their paper caused some stir among the community of physical organic and computational chemists, the situation was well understood by

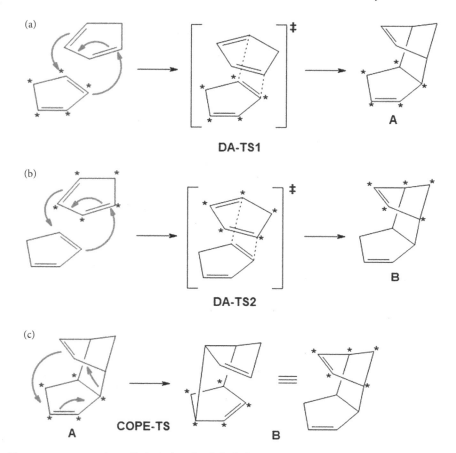

**Figure 18.6.** Reaction of labeled and unlabeled cyclopentadiene. (a) Labeled cyclopentadiene is the dienophile. (b) Labeled cyclopentadiene is the diene. (c) The Cope rearrangement interconverting **A** and **B**.

geometers. Let's look at the surface of Figure 18.8. Starting at the pass ($TS_1$), the black line traces the steepest descent forward. Normally, that would head to a product, but here it connects to a second transition state $TS_2$. $TS_2$ is the transition state for interconversion of the two products, $P_1$ and $P_2$.

Let's follow that black path. At **TS1**, moving forward or backwards means going downhill. That defines the easiest way that takes us over the pass on to the other side. (Or chemically, it defines the reaction coordinate from the reactant through the transition state and on to product.) In the neighborhood around **TS1**, moving off that black path means going uphill. The black path lies in a valley separated by the hills to the left and right. As we approach **TS2** following the black path, we continue to move downhill. However, as we look

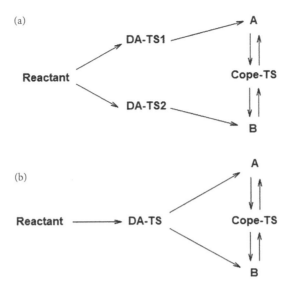

**Figure 18.7.** (a) Classical mechanism for one reactant leading to two products and (b) alternative bispericyclic mechanism.

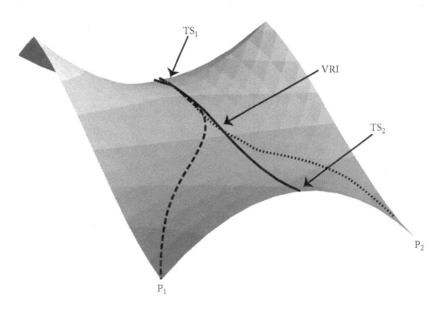

**Figure 18.8.** A surface characterized by a valley-ridge inflection (VRI) point.

to the right or left, we see a dropoff. Instead of being in a valley, we are now on top of a ridge, where moving left or right means falling downhill.

Walking along this black path means transitioning from being in a valley to being on top of a ridge. We passed through the valley-ridge inflection point (VRI), the point where the valley changes to a ridge. And really, once we hit that VRI point, the best thing to do to reach lower-energy regions is to divert our movement off that black path. We can move either to the left or to the right. That's what is depicted by the dashed path that leads to $P_1$ and the dotted path that leads to $P_2$.

This situation is what is found for the reaction shown in Figure 18.5. The two reactants come near each other and pass through a single transition state. Then in the region near the valley-ridge inflection point, the paths diverge (or bifurcate) toward each of the two products. The two products can then interconvert through the Cope rearrangement. This mechanism is depicted in Figure 18.7b.

Caramella called this a *bispericyclic transition state*, but Ken Houk later coined the name *ambimodal transition state* to more generally cover any case where a single transition state leads to multiple different products. Sometimes, these are also referred to as *posttransition state bifurcations*, reflecting the idea that after passing through the transition state, possible trajectories separate and end at different products.

Daniel Singleton, an American chemist, is a leader in identifying reactions with this unusual potential energy surface. A distinguishing characteristic of Singleton's work is his combination of careful experiments coupled with quantum mechanical computations, providing both real data and an interpretation grounded in theory. I provide three examples of Singleton's work here. Don't get caught up in interpreting all of the chemistry of the reactions. Instead, focus on the fact that in each example, *there is only one transition state, even though there are two products.*

The first example builds off of Caramella's bispericyclic theoretical work by examining a reaction where either reactant, **1** or **2**, can be the diene or dienophile in a Diels-Alder reaction (Figure 18.9). The experiment observation is that **3** is made about 2.5 times more than is **4**.

The traditional interpretation of this result would be that the barrier toward forming **3** is lower than that forming **4**. However, the quantum mechanical computations indicate that there is only one transition state and that the minimum energy pathway leads to **3**. To deal with this situation, Singleton was forced to compute a number of trajectories, following the motion of reactants up over the transition state and on to product. This is a much more time-consuming and difficult computation than simply identifying which is the

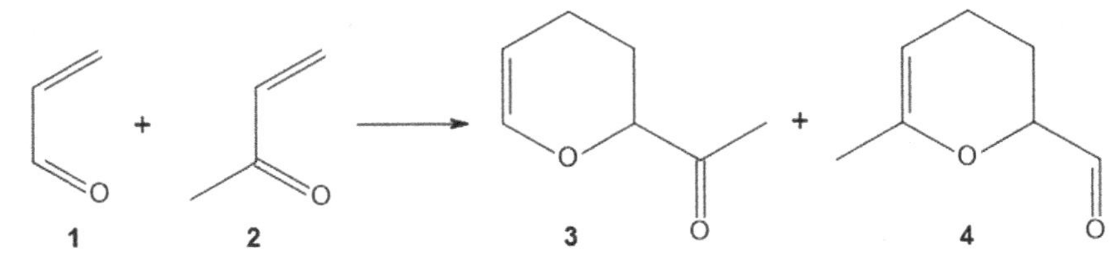

Figure 18.9. First example by Singleton of a bispericyclic mechanism.

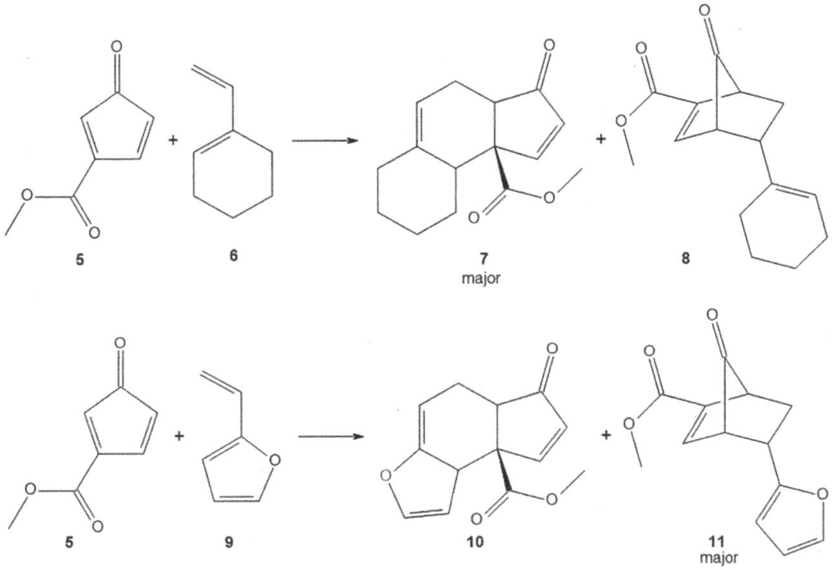

Figure 18.10. Second example by Singleton of a bispericyclic mechanism.

lowest energy transition state. The computation of these trajectories is called molecular dynamics, and it tracks both the position of all of the atoms and their momenta along a reaction. In this case, Singleton computed 296 trajectories, and the number of trajectories that end with 3 is about 2.5 times the number that end at 4, in really remarkable agreement with experiments!

The second example also involves Diels-Alder reactions. The experiments indicate that in the first reaction of Figure 18.10, only product 7 is observed, but in the second reaction, both products are observed, with 11 produced a bit more than is 10. Note that in the first reaction, reactant 5 is the dieonophile in making the major product (7), but in the second reaction it acts as the diene in making the major product (11).

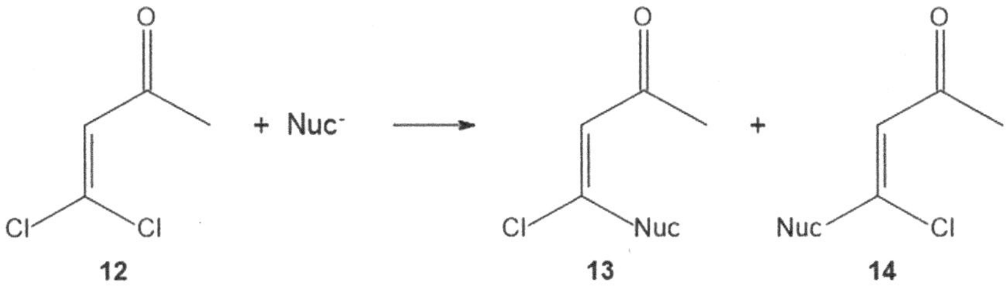

**Figure 18.11.** Posttransition state bifurcation example from Singleton.

For both of these reactions in Figure 18.10, quantum mechanics computations identify only one transition state, and it is bispericyclic. Trajectory computations find that for the first reaction most trajectories lead to **7** and that the low barrier for conversion of **8** to **7** (a Cope rearrangement) results in only **7** being observed. However, for the second reaction, a few more trajectories lead to **11** than to **10**, again consistent with the experiment.

The third example is the nucleophilic substitution reaction shown in Figure 18.11. The experiment indicates that four times the amount of **13** is made than **14**. A number of transition states were located by quantum mechanics computations, but none of them connects to **14**. Traditional statistical theories would then suggest that only **13** would be made, inconsistent with the experiment. Once again, molecular dynamics are needed to understand this reaction. Singleton's computational results are that about 84% of the trajectories end at **13** and 16% end with **14**, consistent with experiments.

These three examples show that posttransition state bifurcations occur in some pretty standard organic reactions. Chemists, especially computational chemists, have identified many more examples, including those involved in the biochemical synthesis of molecules found within cells. Ken Houk has even found a case where a single transition state leads to three different products: a *trispericyclic* example.

Next I discuss another set of reactions that share a different sort of potential energy feature that challenges our previously held perspective. A *caldera* is the technical name for the crater formed when the dome of a volcano collapses. An idealized version is shown in Figure 18.12. An excellent example of a caldera is Crater Lake in Oregon, where the top of the mountain blew off in a tremendous eruption. What was left of the mountain was a crater that filled up with rain to create the lake. A subsequent much smaller eruption created Wizard Island, which has a crater, or caldera, at its top. Visitors standing at the rim might think that Crater Lake gets its name from that small caldera on the

**Figure 18.12.** An idealized caldera.

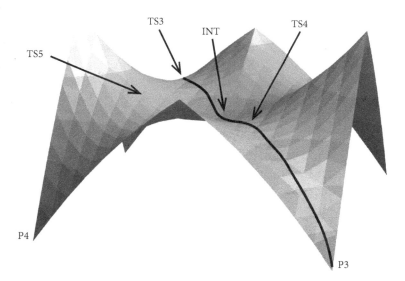

**Figure 18.13.** A sample trajectory on a caldera surface.

island, not realizing at first that they are standing on the rim and looking into a much larger caldera.

Imagine that you are sitting in a sled at the point labeled **TS3** on the ice surface that is shaped like that in Figure 18.13. With a gentle push forward, your sled moves down into the caldera, through the region labeled **INT**. Since **TS4** is the same height as **TS3**, your sled can make it up and over this hill, and then travel downhill to **P3**.

Your initial momentum, coming from the gentle push off at the top of TS3, dictates that you must end up in P3. But what if you want to get to P4? Somehow you have to get over the hill associated with TS5. That requires a right-hand turn, and on an essentially frictionless surface, you can't make that turn. Instead, you will have to provide some initial momentum in the direction of TS5. Another option might be to start in the direction of the hill opposite TS5 and partially climb it and come back down as if you bounced off the hill, now in the direction of TS5. Both of these paths will require somegood luck, pushing off in the right direction with the right speed to finesse the surface in just the right way.

The key feature here is that the caldera is shallow, and the initial momentum will likely just carry you forward. Even if other transition states out of the caldera are lower in energy, and that's the case here—TS5 is a lower barrier than TS4—without some force to push you into that right-hand turn, odds are extremely likely that you will repeatedly end up at P3.

Molecules reacting on a caldera-like surface will behave in this same way. Their initial momentum into the caldera will direct them to cross directly across the shallow bowl and exit straight out. Even if there are alternate exit channels that have lower barriers or that lead to lower-energy products, their momenta will keep them from making the turns needed to access these alternate routes.

In standard statistical theories, molecules will enter the caldera and bounce around inside it for some time, equilibrating all of its vibrations and rotations, and then mostly exit over the lowest barrier. But if the caldera is shallow, this won't occur. Instead, the molecules will spend little time in the caldera, and we will see reaction products determined by dynamics (momentum) rather than statistics.

The best examples of reactions on a caldera-like surface involve rearrangements of small molecules. I present two of them here. The first is the rearrangement of vinylcyclopropane to cyclopentene (15 → 16), shown as the top reaction in Figure 18.14. This appears to be a pericyclic reaction, and numbering the carbon atoms in the figure helps us see which bonds are made and broken in the reaction. The American chemist John Baldwin did a very careful isotopic labeling study of the reaction of 17. If the reaction follows the Woodward-Hoffman allowed pathway, then only 18a and 18b should be made. However, the amount of 18c and 18d observed is nearly equal to the amount of 18a and 18b. Quantum mechanical computations show a caldera-like surface. Trajectory analysis shows a bimodal distribution. Over short periods of time, the allowed products are formed almost exclusively, but the longer trajectories favor the forbidden product. Reaction dynamics are clearly important in this reaction.

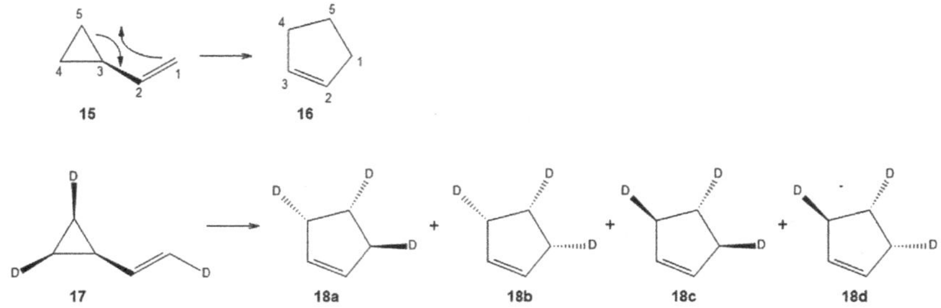

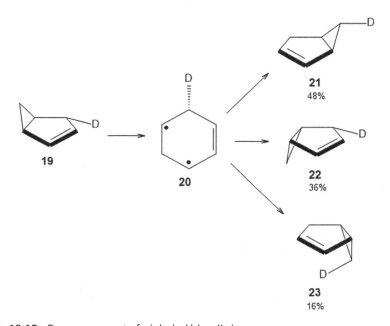

Figure 18.14.  Rearrangement of vinylcyclopropane to cyclopentene.

Figure 18.15.  Rearrangement of a labeled bicyclic hexane.

The second example, a reaction that again appears to be pericyclic, also comes from the work of John Baldwin. **19** can rearrange in a concerted way, but three different products are observed **21–23** (Figure 18.15). Computations implicate an intermediate on the reaction surface, the diradical **20**. A diradical is a reactive molecule in which two carbon atoms have seven electrons, rather than the desired octet. In this case, **20** is formed by breaking the bond shared by the three- and five-membered rings, with the two electrons in the breaking bond moving one to each of the carbon atoms of that former bond.

The diradical intermediate exists in a shallow caldera. Molecular dynamics computations indicate that most trajectories travel straight across the caldera, with no appreciable lifetime of the diradical. The initial momentum carries the molecule across the caldera; the direction of that initial momentum determines the product that is formed. The size of the barriers exiting the caldera play essentially no role, and so once again, a statistical treatment fails to explain the results. Instead, we need to know the shape of the potential energy surface *and* the momenta of all of the reactive species in order to model this reaction.

The examples presented in this chapter give but a taste of the ever-growing list of reactions that display nonstatistical outcomes. There are examples of reactions where the reactant species hover near each other for some time before reacting; nucleophilic substitution like an $S_N2$ reaction except that the nucleophile bumps into the substrate twice before reacting; or reactions that avoid very stable intermediates. Many examples of bifurcating surfaces have been found in the biosynthesis of terpenes, a family of important natural materials within plants and animals.

For most cases where reaction dynamics come into play, the reaction rate is faster than that predicted by statistical theories. In these cases, trajectories often traverse intermediates in a single pass, spending little appreciable time in the energy well associated with that intermediate. There is often an accompanying specificity in the product selection, with preference given to the product that is along the direct path, requiring minimal turning on the potential energy surface.

But there are some cases where reactions can be slower than expected due to dynamic effects. An assumption in statistical theories is that once the molecules on a potential energy surface cross the transition state, they proceed onward toward products; they do not turn around and recross the barrier to go back to reactants.

A nice analogy can be made to the Continental Divide, the mountain ridge that separates the western part of the United States from the eastern part. Rain that falls to the east of the Continental Divide will eventually end up in the Atlantic Ocean, while rain that falls to the west of the Continental Divide will end up in the Pacific Ocean. Water does not cross over the Continental Divide naturally. (Humans have engineered some water systems using pipes and pumps to move water across or under the Continental Divide—but that has required significant expenditure of energy)

In most studies of reactions that display nonstatistical dynamics, a few, often many, trajectories are found to recross the transition state barrier. The trajectory starts at the reactant, climbs over the barrier, and then bounces off some potential energy wall that redirects the molecule back over the barrier to reactant. This is obviously a nonproductive trajectory, at least over some short time. That implies that fewer reactions take place than what statistical theories suggest because the statistical theories all assume that, once over the barrier, the reaction will go onward to product, never to recross the barrier. It's probably fair to say that few chemists were thinking that recrossing a transition state was a frequent occurrence!

Physicists have known for centuries that knowledge of both position and momentum (i.e., phase space) are necessary to describe and predict the motions of any system, whether that be planets around the sun, or billiard balls on a pool table. So, it came as no surprise to some chemists that we too would need to consider position and momentum. However, what has been a surprise is the extent to which nonstatistical dynamics are being observed. For decades, our use of statistical theories with a focus on just the potential energy surface, and often just local energy minima and transition states, had been sufficient! Perhaps we got a bit complacent. The success of so many reaction mechanisms, especially the broad applicability of orbital symmetry rules, was so pervasive that we fell for our simple rules and models and gave no thought to what role momentum might play.

The many studies over the past two decades that have exposed the important role of reaction dynamics in many seemingly well-understood reactions set the stage for creating a new paradigm. The ever-increasing number of reactions that do not obey statistical dynamics tell us of the limits of our reaction mechanism model. That's a good thing; it is a standard part of how science works. New ideas, new experiments, and new models are generated by pushing at the limits of our current models.

It is somewhat unsettling, though, since the current situation leaves us in limbo. We recognize from these studies that our notion of the reaction mechanism has serious limitations, but as yet we have no viable replacement model. We have little guidance as to when our statistical theories might fail. When they do fail, our recourse is to adopt molecular dynamics computations, computations that are not routine and require significant resources. Researchers must learn to be on notice for experimental and computational hints suggesting that statistical treatment may not be applicable, and should be ready to apply trajectory analysis to their problem.

This situation is also a serious challenge for organic chemists who are educators. How do we present nonstatistical dynamics to students who are just learning the field? Coming to grips with reaction mechanisms that follow statistical theories is challenging enough, though I hope I have made this topic less mysterious in this book. Nonstatistical dynamics is even more abstract and will be a challenge for any beginning student to embrace, especially since we can't provide them with any real organizing principle. All of our introductory textbooks omit this topic, and even textbooks for advanced and graduate courses have largely avoided nonstatistic dynamics. Developing good pedagogies for reactions that display nonstatistical dynamics is the great challenge facing physical organic chemists today!

I am actually very excited by these developments. One might think that the lack of a unifying theory means confusion and uncertainty—and there is certainly plenty of ennui among those, dealing with this topic. But it has been very exciting to watch these early days, reading the new papers, attending conferences and arguing with colleagues over what this all means and over how we should move forward. It's exciting to witness, and perhaps participate in, the development of a new paradigm.

I have always been jealous of the young physicists who went through graduate school or postdoctoral training in the early 1920s. They were first-hand witnesses to the birth of quantum mechanics. The field was initially at a great loss for ideas. The world seemed to be shifting upon an unstable landscape. New experiments were reported with strange results. Explanations were offered but then shot down for being incomplete or lacking mathematical rigor. What must it have been like to be part of Niels Bohr's inner circle at Copenhagen or among Max Born's school at Göttingen, knowing that any day someone might have the piercing insight to change the world of physics? What an opportunity to contribute to a brave new world!

I hope that the next decade will elicit the same sort of excitement and change in my discipline of physical organic chemistry. While today we struggle to make sense of nonstatistical dynamics, I have faith that soon my colleagues will create a new paradigm, a new model, to represent organic reactions more accurately. We will surely witness an explosion of new ideas and greater understanding of organic chemistry, with far-reaching consequences and impacts.

# 19

# Lessons Learned

I have only scratched the surface of the intricacy and beauty of organic chemistry. A glance at recent Nobel Prizes in Chemistry, as follows, testifies to the continuing excitement surrounding organic chemistry. (Don't be concerned about whether you understand the chemistry for which these prizes were recognized. The work here is technical. Rather, the point is that organic chemistry is a vibrant, active science.)

2022: Carolyn R. Bertozzi, Morton Meldal, and K. Barry Sharpless received the Nobel Prize in recognition of their work in click chemistry and biorthogonal chemistry. Click chemistry is the idea of taking simple, readily available reagents that react quickly and economically—they just "click" into place. A key idea here is that two reactants are specific for only each other, and their reaction proceeds without need for a catalyst or other aid. The biorthogonal extension is to apply click chemistry of biologically active molecules, without affecting their biological activity. This is Sharpless' second Nobel Prize; his first was in 2001, as described below.

2021: Benjamin List and David Macmillan were awarded the Nobel Prize for their development of asymmetric organocatalysis. I've mentioned catalysts a few times before; organocatalysis makes use of organic molecules as the catalytic agent. This is important in that most catalysts are metal-based, as you will see in many of the other Nobel Prize examples below. The term *asymmetric* used here means that the reactions produce one enantiomer preferentially over the other. List and Macmillan's asymmetric organocatalysis was inspired in part by how some enzymes work within the cell.

2016: Jean-Pierre Sauvage, J. Fraser Stoddart, and Bernard L. Feringa developed molecular machines. The small molecules designed and synthesized by these chemists mimic real machines: tiny rotors, motors, and molecules that capture and hold other molecules.

2010: Richard F. Heck, Ei-ichi Negishi, and Akira Suzuki were recognized for their work in developing metal-catalyzed coupling reactions in organic synthesis.

*Thinking Like a Physical Organic Chemist.* Steven M. Bachrach, Oxford University Press. © Oxford University Press 2023.
DOI: 10.1093/oso/9780197640371.003.0019

Coupling reactions join two organic molecules together. This technology has seen incredible application especially to attach substituent groups to aromatic rings.

2005: Yves Chauvin, Robert H. Grubbs, and Richard R. Schrock were recognized for their development of metathesis reactions. Metathesis reactions of the type developed by these three chemists utilize a metal to bring together two carbon–carbon double bonds and then swap their carbon atoms along with the groups attached to them.

2001: William S. Knowles, Ryoji Noyori, and K. Barry Sharpless were awarded the Nobel Prize for their work in catalytic hydrogenation. Again, the catalysts involve a metal atom. Hydrogenation is an addition reaction in which water is added across a double bond: a hydrogen atom attaches to one carbon, and the OH group adds to the second carbon. Sharpless's contribution is of particular note as his catalytic system creates one enantiomer in high preference to the other one.

Most of these examples provide new methods for synthesizing molecules, allowing for more efficient production of pharmaceuticals, food additives, pesticides, and precursors to ceramics and polymers. These examples date back just twenty years; significant organic chemistry developments were behind many of the biochemistry-based and medicine-based awards in this period.

Organic chemists continue to develop new methods to create compounds, along with new molecules with desired properties and chemical behaviors. Physical organic chemists' efforts to understand how reactions occur continue unabated as well, as they focus on developing a deeper understanding of why molecules behave the way they do and then what might be done to push molecules along alternative reaction paths.

So, what can we take away from the examples provided in this book? How can the approaches to problem-solving developed and honed by physical organic chemists apply to other areas? Can this problem-solving approach assist scientists in other disciplines or perhaps people active in areas removed from the sciences? Can these methods work in the social sciences, in political science, or even in the arts and humanities?

I don't claim that physical organic chemists have some magical processes, some special sauce that especially distinguishes this discipline. In fact, most of the following discussion is part of good scientific practice that can be found anywhere from an astrophysics lab to a zoology lab. Nonetheless, I do believe that the examples from my discipline ably demonstrate how the best aspects of the scientific method help advance our field, especially in providing guidance in evaluating the works of others.

Following are a few key elements for thinking about a problem, approaching a solution, and warding off biases that may keep you from identifying an original solution.

## 19.1. Allegiance to Fact

The bedrock of science is fact. A fact is an observation that is reproduced and confirmed by others. In chemistry, a fact might be the boiling point of a compound, the elemental composition of a compound, the rate of a reaction, or the amount of heat released when a reaction occurs. An observation becomes fact when it's been reproduced, preferably by an independent scientist at a different location.

A fact is the result of a well-designed experiment that is often built to probe a hypothesis. Facts inspire hypotheses, an explanation for those particular observations. The collection of facts can lead to scientific laws and theories. In the absence of facts—facts recognized by practitioners across the globe—science cannot be performed and scientific progress cannot take place.

In the context of this book, it should be clear that a reaction mechanism is *not a fact*! Facts are collected to guide us toward developing the reaction mechanism. Any proposed mechanism that is not consistent with all of the facts is wrong and should be utilized with extreme caution, if not outright discarded. Furthermore, the results described in the last chapter call into question the fundamental meaning and value of the mechanism itself.

It seems almost tragicomic that this notion of the centrality of *fact* within science needs to be discussed at all. Science must be grounded in observation, in experiment. There is room for discussion and disagreement in the *interpretation* of facts—but the facts themselves, when properly vetted and reproduced, are not subject to disagreement. Water boils at 100°C at sea level and at lower temperatures as one moves to higher elevations; that's not up for debate. If you wish to offer an alternative explanation to the commonly held explanation (that centers on vapor pressure), good scientists will listen and question.

However, in 2022 we live in an Orwellian world where facts are fungible, where "alternative facts" can be offered in total seriousness, where individuals appear free to choose or create their own "facts." Somehow, society has allowed the sanctity of fact to erode, to become nothing more valuable than a personal belief.

Facts need to cross national borders and political divisions. For science to continue to offer explanations of the world that lead to technological and engineering solutions to our problems, that make the world a safer, healthier, and more enjoyable place, we all must work to ensure that, if nothing else of the scientific process is preserved, this notion of the universality of a fact must be firmly reinstated. Scientists need to proactively show allegiance to facts.

## 19.2. Careful Development of Hypotheses

Coming up with a hypothesis is the first step of any scientific inquiry. The hypothesis proposes some explanation or question about some fact or set of facts. The hypothesis guides the scientist toward deciding on the next experiments to run.

A good hypothesis should be narrowly focused. If the hypothesis is too broad in scope, then designing an experiment that can address the hypothesis will be difficult, if not impossible. Experiments are best if they answer a very specific question, addressing one variable at a time. Experiments that have multiple variables are often difficult to interpret.

Suppose you are having difficulties connecting to a Zoom conference. You think that perhaps the computer is just lost and that rebooting it will solve your problem. You have applied this solution in the past with some success, so your hypothesis has some foundation. So you reboot the computer.

Two outcomes are now possible. First, after rebooting, you connect to your Zoom call and all proceeds well. For the immediate problem, it appears that your hypothesis has been corroborated. However, what have you really learned? What was the underlying cause of your inability to connect to your meeting? Was the WiFi working? Did you have the correct link? Did you call in too early? Do you have an outdated version of the app that is known to be unstable? Did you have too many other applications open at the same time? Rebooting may have solved your issue for the moment, but it provides no real guidance for preventing a similar occurrence from happening tomorrow.

Suppose the second outcome happens, and after rebooting, you still can't connect to the call. What have you learned that will help you solve the problem? What do you do now? You think your meeting has already started, and you have no new operational information to help you connect. In fact, you've learned very little because your hypothesis was so broad. Instead, asking simpler questions can help you step through single variables to identify the culprit. Is the WiFi working? Do I have the right time for this meeting? Can I upgrade the Zoom app?

A good hypothesis must suggest an experiment to run in order to test the hypothesis. This goes back to Karl Popper's notion of falsifiability. If one can't obtain some data to help verify the hypothesis, then the hypothesis rests outside of science. Whether a hypothesis that can't be tested has some value to you rests on factors outside of science, such as belief or faith. Within the practice of science, experiment must be the final arbiter of truth.

## 19.3. Careful Experimental Design

A careful experiment is one for which we have confidence in the result, and the result provides some insight toward addressing our hypothesis.

What are some components of a carefully designed experiment? I'll discuss this issue by introducing a hypothesis and talking through the experimental design elements. Our hypothesis will be that talking on a cell phone distracts the driver enough to be dangerous. We need to figure out some test that will evaluate "distraction." We might decide to create a driving course defined by cones. The course will have some left- and right-hand turns, some stop signs and stoplights, and a parallel parking exercise. We will evaluate each driver's time through the course along with how many cones are knocked over. We might expect that a distracted driver will take longer to complete the course and will knock over more cones than an undistracted driver will when driving that same course.

We recognize that individuals will approach the course with their own biases—some people are by nature more aggressive, some more reserved. So, we will have each driver do the course twice, once while holding a cell phone and conversing and once without the phone, and we'll measure the difference in travel times and cones hit between the two trials. Since a driver will have some added knowledge of the course the second time through, each driver will flip a coin to decide whether their first run will be with or without the phone.

The pass through the course without any distractions is our control study. A control removes all of the variables that we think might affect the outcome, providing a baseline comparison. We can further classify this as a negative control: a control without any of the presumed effects in place.

We might consider including a positive control, an experiment in which we know we will observe the effect. In our case, a positive control would be to have a driver run the course while truly being distracted. For example, we might put a video screen near the dash and ask the driver to report back the synopsis of the movie. Or we might ask the driver to hold a hot cup of coffee in one hand while maneuvering through the course. These positive controls provide us with data on how poorly a truly distracted driver will do driving the course.

We have, say, thirty drivers traverse the course, once with a cell phone and once without. We have half of the drivers take a third pass, this time with a significant distraction affecting their performance. We note that the cell phone passes take longer, with more cones hit than in the control runs. And the seriously distracted drivers take out almost every cone. At this point, we might feel pretty confident that we have a set of tests that will provides us with good data that supports our hypothesis.

A good scientist will now consider whether the experiments might be biased in some way. Are there other factors at play here that we neglected to consider? Are other explanations possible?

We might consider just what it is about using the cell phone that causes the distraction. Is it holding the phone that ties up one of our hands? Or is it our concentration on the conversation? We might then have some drivers run the course just holding the phone but not talking. Other drivers might run the course talking in hands-free mode. We might have drivers run the course with a passenger to see if conversing in person or via phone makes a difference. We might have some drivers run the course while listening to a podcast.

Perhaps it would be worth considering how much attention each driver paid to the conversations or stories. We might follow each drive with a brief questionnaire to assess how much each driver retained from the conversation or story told during the drive about the course.

It should be clear that every experiment leads to additional questions, new hypotheses, and better refinement of our understanding of the problem at hand. This continued series of refinement of experiments is what science is all about. It also points to the never-ending quest that is science—there is always another question that can be asked. Science doesn't come to an end; rather we incrementally build up a set of laws and theories and models that help us predict the outcome of experiments with improving accuracy.

Another key element of good experimental design is one I have mentioned a few times in the book. When possible, design an experiment whereby the effect you believe is dominant is set up to fail. If, even under these difficult circumstances, the effect still plays out, you have extremely strong support for that hypothesis.

Perhaps this notion is best understood through examples. Consider the endocyclic restriction test that I discussed in Chapter 9. Eschenmoser's operating hypothesis was that the $S_N2$ reaction proceeds through a backside attack that needs to be essentially linear. His experimental design (see

Figures 9.7–9.10) allows for a linear backside attack only if two molecules come together to react. An attack angle of about 120° is a possible alternative through a unimolecular process. Entropy dramatically favors that unimolecular process over the bimolecular process. This situation disfavors the effect his hypothesis proposes. Nonetheless, linear backside attack is so critical that it overcomes the entropic barrier designed into the experiment.

The second example comes from Cram's experiment testing the hypothesis of anti-periplanar elimination (Figure 10.9). Cram designed a molecule which, should it react by his hypothesized pathway, would produce a product less stable than if the reaction proceeded by a different pathway. Again, the experimental design is set up to disfavor the effect (anti-periplanar elimination) that he proposed. Nonetheless, the less stable product is the only one observed.

Both of these experiments resulted in observations that strongly supported the proposed hypotheses. These valuable observations testify to these chemists' clever thinking. This lesson is one that can be employed throughout science and elsewhere.

## 19.4. Look to Adjacent Disciplines for New Techniques and New Ideas

I believe that a major lesson from the practice of physical organic chemists is our community's enthusiastic embrace of new ideas and new technologies from other disciplines. We are happy, if not eager, to exploit new developments from other disciplines and relate them to our own set of problems.

Given the qualifier "physical" to our discipline name, it may not be surprising how readily we have taken developments in physics and applied them to organic chemistry. I have already discussed how Hückel applied quantum mechanics to aromatic systems, followed soon thereafter by Linus Pauling's development of resonance. As computers were developed in the 1960s and 1970s, computational chemistry exploded with the application of quantum mechanics to ever-larger organic molecules. This culminated in the award of the 1998 Nobel Prize in Chemistry to John Pople for his development of methodology and programs for computing the properties of molecules and to Walter Kohn for his development of density functional theory, a revamping of the mathematical application of quantum mechanics that has dramatically reduced the computational size, enabling study of much larger molecules. Computational organic chemistry has become its own fully realized subdiscipline and a major asset to organic chemists of all stripes to address their projects.

Perhaps a more compelling story of the transition of technology from one scientific discipline to another is contained in the history of nuclear magnetic resonance (NMR). NMR relies on nuclear spin, the property where some nuclei behave like a magnet. When placed inside a large external magnet, these nuclear spins will align either parallel or antiparallel to that external magnet. The nuclear spin can then be "flipped" by exposing the nuclei to radio waves. This observation was first made by Isador Rabi for gas-phase atoms, and almost a decade later Felix Bloch and Edward Mills Purcell observed this spin flip in solids and liquids. All three of these physicists earned the Nobel Prize in Physics: Rabi in 1944, while Bloch and Purcell shared the 1952 prize.

The hydrogen nucleus has spin, and it was soon observed that the hydrogens in a molecule "flip" with different frequencies of radio waves. This is analogous to the different frequencies of radio stations: your favorite oldies station is found at a different location on the dial than the power pop hits station or your local news station. You need to tune the dial to match your receiver to the frequency of that station; just a little off and the reception is filled with static. Similarly, different hydrogen atoms will flip with a different frequency, and we will need to dial in that frequency to get it just right. Physical organic chemists soon correlated these NMR frequencies with different chemical environments, such as hydrogen on a primary or secondary carbon, hydrogen attached to a carbon in a double bond, or hydrogen attached to an aromatic ring. Knowing these correlations, and some additional features of the signals, chemists can deduce the structure of a molecule. NMR is the go-to resource for determining chemical structures; NMR is extremely sensitive to small structural changes, requires only a small sample to perform the experiment, and the sample is not damaged in any way during the experiment. NMR is ubiquitous through organic and inorganic chemistry. This technique is introduced in the first-year organic chemistry sequence, and these novice students learn to identify unknown samples purely through their NMR spectra.

NMR experiments expanded to probe spin flip of different elements at the same time, such as simultaneously examining the flip of hydrogen and carbon nuclei—which is sort of like tuning into the AM and FM bands at the same time. This advance led to the ability to determine the structure of much more complex molecules, like proteins, and even to obtain information on the conformation of molecules. With the use of variable temperature NMR, physical organic chemists are able to obtain kinetic information, such as the rate of chair ring flip of six-membered rings. This technique can be extended to determine the energy of chemical reactions.

That's not the end of the story for NMR. Since hydrogen and carbon atoms are found in living species, NMR was adapted to probe the interior of animals too. We call this magnetic resonance imaging (MRI), but it really is just the NMR of living tissues, with complex imaging software interpreting the frequencies of the spin flips. MRI is now a standard tool in medicine for examining soft tissue, including ligament tears, cancers, and spinal disk injuries. As with organic compounds, a major benefit of MRI is its sensitivity to small changes in the environment, for example, allowing for readily discerning healthy tissue from cancerous cells. But perhaps its best attribute is that it is completely harmless. Unlike x-rays, which expose the patient to radiation that can damage healthy tissue, MRI exposes the patient to radio waves that we are subjected to all the time without any harm.

MRI has now transitioned from the medical field to neurology and psychology in the form of functional-MRI (fMRI). I mentioned this technique in Chapter 8, whereby the fMRI experiment detects activity in the brain. Researchers are now using this method to understand how the healthy brain functions and how to detect abnormalities without dangerous invasive surgeries. Back in the 1940s, Rabi, Bloch, or Purcell likely had no notion of the potential future applicability of their work. So NMR has moved from physics to chemistry to biochemistry to medicine to psychology. What's next?

This story of NMR's progression from discipline to discipline is not unique in history. Leonardo da Vinci and James Audubon applied their artistic skills to scientific pursuits. In the late nineteenth and early twentieth centuries, a transfer with some serious negative consequences was made by the social scientists who co-opted Charles Darwin's notions of natural selection and applied them to social arenas through now-discredited ideas such as Scial Darwinism and eugenics. In addition, Einstein's theory of special and general relativity influenced the novels of Virginia Woolf, Thomas Pynchon, Milorad Pavic, and Joseph Heller.

Scientists should always be open to the possibilities of some new development, some new theory, some new experiment that can be repurposed for their own questions. New techniques and new theories open up possibilities that were previously unimagined even by the developers of that new technique. This transfer of intellectual property from one discipline to another acts like new fertilizer; it can grow hypotheses never before imagined.

## 19.5. Embrace Beauty

In Chapter 8, I mentioned the important role beauty has played in inspiring scientists. Beauty comes in many forms. I think for scientists, beauty is often reflected in symmetry. Symmetry has inspired the synthesis of many of the molecules I have discussed, such as the Platonic solids and starphenylene. The symmetry of fullerene certainly played a major role in how this molecule captured the imagination of so many chemists when it was first discovered.

I have also discussed the beauty of mathematical formulas: the elegance of an equation that is compact or that involves a few fundamental constants is undeniable. Scientific theories are also beautiful when they succinctly capture the nature of a broad range of observations or when they open up new vistas upon the universe.

Let beauty be a guide in your scientific endeavors. Follow symmetries, for they will often guide you to insight. Be open to opportunities to break symmetry, to observe what results from near alignments. Symmetry breaking is often associated with important physical attributes. For example, the discovery of charge violation—that some physical processes are not the same when matter is swapped with antimatter; parity violation—that some physical processes are not the same when reflected through a point; and time violation—that some physical processes are not the same when run forward and backward—all led to new theories. This ultimately resulted in new symmetry: the CPT Theorem stating that charge, parity and time are collectively preserved.

The 2004 Physics Nobel Laureate David Gross nicely summarized the role that beauty and symmetry may play in scientific pursuits:

> At the fundamental level nature, for whatever reason, prefers beauty and is marvelously inventive in inventing new forms of beauty. If this is the case then it provides us with an important tool for the exploration of nature. When searching for new and more fundamental laws of nature we should search for new symmetries. (David Gross, *Proceedings of the National Academy of Sciences USA*, 1996, *93*, 14256)

## 19.6. Question Underlying Assumptions, Especially Hidden Assumptions

Everyone carries biases, regardless of the situation. Your training and education, your experiences, your interactions with people, the works you've read, and the media you've consumed all influence how you approach a new problem. These biases undergird the assumptions you hold as you look at some new data and decide what to do next.

At all stages of the scientific process, one needs to be cognizant of the assumptions that have been made. Assumptions can lead you to ask a certain question or to make a particular interpretation. Often these assumptions are critical to your process: they can provide guardrails that will help you to avoid straying too far from standard practices.

At the same time, assumptions may keep one from asking the probing question or gaining the proper insight. Your assumptions might be so strong that you can't even conceive of any alternatives, and so you don't ask the right question, you don't perform the right experiment, you don't see the alternative explanation.

A frequent occurrence in a synthetic chemistry laboratory is for a new reaction to produce black goo at the bottom of the flask. Most often, we simply complain that the reaction has gone to tar, and we complain about the time wasted in conducting that reaction. We simply clean the flask and disposing of that tar. Sometimes, though, that black goo turns out to be a very interesting product, one that was unanticipated or one that provides some interesting view into what's happening during the reaction. Questioning the assumption that the expected product would not be black goo, that black goo is typically some plastic-like compound that results from too much heat, may be just what is needed to make that coveted breakthrough.

I have discussed a number of experiments that were expressly designed to question assumptions. Eschenmoser's brilliant endocyclic restriction test (Chapter 9) addressed the assumption of linear backside attack, which turned out to be a valid assumption. Schleyer probed whether backside attack is truly impossible in a tertiary system (Chapter 12), an assumption he found to be false. The gas-phase study by Brauman questioned whether alkyl groups are electron donors (Chapter 12), an assumption his experiments showed to be false. These and many other experiments became classics in physical organic chemistry because of the elegance of their experimental design tackling such important, seemingly foundational, notions of organic chemistry.

The lesson is to be unafraid of taking a step back to think about what you are presupposing before you begin that next experiment. More often than not, your assumptions are likely to be well founded. They've become part of your scientific intuition to help you filter the signal from the noise. Occasionally, however, your assumptions are misguided and can keep you from running the proper experiment or considering that alternative interpretation that will reveal that what we have long understood to be part of our standard model is in fact erroneous. Undeniably, these situations disrupt our equilibrium and lead to major breakthroughs in science.

You need to be especially on guard for hidden assumptions. Hidden assumptions—those that you are unaware you have even invoked—typically are so fundamental that one may perhaps think of them as facts. That's why they can prove to be insidious—you are not even aware that you have assumed some concept that might be steering you in the wrong direction.

## 19.7. Embrace Change

I think there is no more important lesson from science than to be prepared for new or revolutionary ideas. Thomas Kuhn, an eminent philosopher of science, argued that science progresses through revolution. Theories become widely embedded within the discipline, perhaps taking on an almost mythical status. Eventually, some experiment or some new interpretation of data is reported that shakes the discipline to its core, and a scientific revolution remakes the discipline.

The classic example of a scientific revolution dates back to the sixteenth century. From the time of the Ancient Greeks, it had been understood that the Earth was at the center of the universe and that all the heavenly bodies circled the earth. Ptolemy had formalized this model to conform to experimental observations of the movements of planets, such that the planets followed paths that were circles within a circle. Nicolaus Copernicus proposed the heliocentric model, with the sun at the center of the universe and the planets, including earth, moving in circular orbits about the sun. It wasn't until Isaac Newton's work that a physical model explained, and provided predictions, of the motion of the planets and other space objects like comets.

The Newtonian view was firmly embedded into physics for centuries. The first cracks appeared when more detailed observations of the movement of Mercury showed its path deviating from that predicted by Newtonian mechanics. Einstein's theory of g relativity provided a new paradigm, ushering in a dramatically different model of the universe, with massive objects curving space itself. The current nascent notions of a dark force and dark matter will likely develop into a new paradigm, yet another cosmological revolution.

In this book I have discussed two revolutions within the field of physical organic chemistry and two areas that are revolutions still in the making. Woodward and Hoffmann's orbital symmetry rules revolutionized how organic chemists viewed mechanisms, moving this intellectual organizing tool into the central position for all of organic chemistry. The discovery of fullerene

(Figure 8.10) presented a new allotrope of carbon. Allotropes are different forms of an element. Carbon is now understood to come in two common allotropes, graphite and diamond, and smaller allotropic forms like fullerenes. The discovery of fullerene opened up a new field of nanotechnology, including such molecules as nanotubes, nanorods, and nanosheets. Many of these molecules have interesting properties and real-world applications in communications devices and monitors.

The two still-developing revolutions in organic chemistry were presented in Chapters 15 and 18. In Chapter 15, I discussed the development of tunneling control as an alternative to the well-established synthesis tools of kinetic control and thermodynamic control. The applicability of tunneling control is at present very limited, and it remains to be seen whether it may be applied to more general cases. Only then might tunneling control be viewed as part of a scientific revolution.

The second revolution-in-the-making is the growing number of reactions that do not follow our statistical theories. Many physical organic chemists recognize the need for some new model to provide predictive guidance as to when dynamic effects will be important. We will also need to develop a model that will allow us to understand the outcome of reactions governed by dynamic effects, preferably with some simple set of rules or back-of-the-envelope calculations. This might lead to some modification of the reaction mechanism model, or perhaps even to its demise. We physical organic chemists certainly have our work cut out for us!

Scientific revolutions create upheaval, but contrary to popular myth, scientists generally have adapted quickly to the changes. Consider the revolution created by the work of Copernicus. The common myth is that his positioning of the sun at the center of the universe was viewed as a demotion of earth and humans, and thus rejected by the Catholic Church as well as scientists. However, as Mano Singham's studies have shown that, most astronomers at the time adopted Copernicus's model because it simplified some computations. Copernicus' work was taught within Catholic universities for decades after its publication, and Copernicus retained his clerical position until he died. It wasn't until more than sixty years following the publication of his work that the Catholic Church condemned heliocentricism. The story of the burning at the stake of Giordano Bruno in 1600 (for reasons of heresy likely unconnected to his belief in the Copernican theory) and the famous trials of Galileo have been exaggerated to tell a mythic story of rejection of science.

Einstein's theory of general relativity was met positively by the physics community. Following Arthur Eddington's experiment measuring the deflection of starlight by the sun, general relativity was broadly accepted, which turned Einstein into a multimedia star. The revolution produced by Watson and Crick's 1953 publication of the structure of DNA was also quickly embraced by the biology community, with the central dogma of molecular biology produced within five years.

The birth of quantum physics is associated with the work of Max Planck, who in 1900 published a paper explaining the theory of blackbody radiation. Blackbody radiation can be pictured as follows. Imagine a hollow metal sphere with a small hole. As this sphere is heated up, it will start to glow and emit light through that small hole, which is called blackbody radiation. Prior to Planck's work, classical physics was unable to properly describe the relationship of temperature to the frequency of the light emitted by the blackbody radiator.

Planck's breakthrough was to suppose that energy is available as small, discrete packages, or quanta, and not as a continuum. It's as if a car can travel at certain set speeds, say 10 miles per hour and 20 miles per hour and 30 miles per hour, but can never travel at 25 miles per hour. Planck viewed this as a mathematical trick providing a mathematical formula that precisely matched experiments. However, he didn't really believe in a quantized world.

Einstein was probably the first physicist to understand and truly believe in quantization. His work on the photoelectron effect in 1905, for which he was awarded the Nobel Prize in Physics in 1921, argued for quantization of light and the energy of electrons in a metal. It wasn't until 1911 that Einstein was able to convince Planck of the reality of quantization. The physics community was then united in an effort to formalize the notion of quanta, which took another fifteen years to complete.

The rejection of scientific revolutions has largely been outside of science. Copernicus' work was adopted by astronomers of his time, but the Church rejected the work on largely political grounds. Darwin's theory of natural selection became the organizing principle within biology but remains contentious outside the discipline.

Anti-Darwinism reached its peak in the Soviet Union with the rise of Lysenkoism in the mid-twentieth century. Trofim Lysenko used his position as director of the Institute of Genetics within the Academy of Sciences to reject natural selection. This politicization of science ultimately led to thousands of biologists losing their positions or even being executed. The effect was devastating to Soviet biology research, from which it has not yet fully recovered. A similar scenario played out in Nazi Germany, with the rejection of relativity and much of quantum physics as "Jewish Science," leading to the loss of hundreds of leading scientists and scholars from German universities. Prior to the rise of Nazism, Germany was the world leader in physics and chemistry. The exile of scientists in the 1930s and 1940s played a major role in the rise of American science and the decline of German science.

I end this book with a cautionary tale. Historically, broad, countrywide political and social rejection of science has invariably led to dark times, with economic downturn and loss of intellectual vigor. People outside of science have feared the potential changes wrought by the scientific revolution—the demotion of the centrality of humans, religion's loss of power, a seeming rejection of divinity—but those fears led to repression and intellectual stagnation.

We have recently witnessed the necessity of multidisciplinary approaches to complex societal problems with the SARS-CoV-2 (Covid-19) pandemic. In about eighteen months, scientists identified the virus, sequenced its genome, developed half a dozen vaccines, including two using new messenger RNA technology, manufactured billions of doses, and vaccinated billions of people worldwide. The speed of the development and distribution of the Covid-19 vaccines is unprecedented in history. It is truly a remarkable scientific, technological, and economic success.

But what is preventing many countries, especially the United States, from reaching vaccination levels of about 90%, where herd immunity likely can be obtained? It's not the technology; plenty of vaccines are available. Rather, the problem lies with politics and sociology and fearmongering and misinformation. We need better leaders and a better-informed population skilled in areas associated with the humanities—communication, media, politics, religion, sociology.

Global climate change requires this same multidimensional approach if we hope to make some real environmental change. If we simply wait for technology to solve our problems, our fate is sealed, and it's not good.

As I am writing this, the 2021 United Nations Climate Change Conference (COP26) has just concluded. The conference brought together leaders and organizations from myriad disciplines and many countries across the world. The conferees endorsed a plan that would take some initial steps toward addressing global climate change. We will see if countries around the world will comply with this plan. But even if they all do, yet more change will be needed to forestall the disasters that will ensue if temperature increases cannot be reduced.

I hope that leaders and citizens of the world can apply some of the lessons and strategies described in this chapter to tackle climate change. We need to question our assumptions, design experiments that provide solid data, seek out opportunities to reach across disciplines, and develop a hypothesis that will lead to both technological and societal solutions.

Most of all we need to *embrace change*. Navigating climate change will necessitate significant alterations to our lifestyles and our economies. We need to see these changes as a continuing part of our human history, which is replete with adaptations to the environment and new technologies. Good scientific practice can aid in this next adaptation. Let's all hope that our political leaders, corporate leaders, scientists, media pundits, film stars, authors, engineers, economists, health professionals, entrepreneurs, musicians—really all of us—can join together to address climate change. The fate of our world is in our collective hands.

# Bibliography

Bachrach, S. M. *Computational Organic Chemistry*. 2nd ed. Hoboken, NJ: John Wiley, 2014.

Baldwin, J. E., and Keliher, E. J. "Activation Parameters for Three Reactions Interconverting Isomeric 4- and 6-Deuteriobicyclo[3.1.0]hex-2-enes." *Journal of the American Chemical Society* 2002, *124*, 380–381.

Baldwin, J. E., Villarica, K. A., Freedberg, D. I., and Anet, F. A. L. "Stereochemistry of the Thermal Isomerization of Vinylcyclopropane to Cyclopentene." *Journal of the American Chemical Society* 1994, *116*, 10845–10846.

Beak, P. "Determinations of Transition-State Geometries by the Endocyclic Restriction Test: Mechanisms of Substitution at Nonstereogenic Atoms." *Accounts of Chemical Research* 1992, *25*, 215–222.

Bogle, X. S., and Singleton, D. A. "Dynamic Origin of the Stereoselectivity of a Nucleophilic Substitution Reaction." *Organic Letters* 2012, *14*, 2528–2531.

Brauman, J. I., and Blair, L. K. "Gas-Phase Acidities of Alcohols. Effects of Alkyl Groups." *Journal of the American Chemical Society* 1968, *90*, 6561–6562.

Caramella, P., Quadrelli, P., and Toma, L. "An Unexpected Bispericyclic Transition Structure Leading to 4+2 and 2+4 Cycloadducts in the Endo Dimerization of Cyclopentadiene." *Journal of the American Chemical Society* 2002, *124*, 1130–1131.

Carroll, F. A. *Perspectives on Structure and Mechanism in Organic Chemistry*. 2nd ed. Hoboken, NJ: John Wiley, 2010.

Cram, D. J., Greene, F. D., and Depuy, C. H. "Studies in Stereochemistry. XXV. Eclipsing Effects in the E2 Reaction." *Journal of the American Chemical Society* 1956, *78*, 790–796.

Diercks, R., and Vollhardt, K. P. C. "Tris(benzocyclobutadieno)benzene, the Triangular [4]phenylene with a Completely Bond-Fixed Cyclohexatriene Ring: Cobalt-Catalyzed Synthesis from Hexaethynylbenzene and Thermal Ring Opening to 1,2:5,6:9,10-tribenzo-3,4,7,8,11,12-hexadehydro[12]annulene." *Journal of the American Chemical Society* 1986, *108*, 3150–3152.

Dobrowolski, M. A., Cyranski, M. K., Merner, B. L., Bodwell, G. J., Wu, J. I., and Schleyer, P. v. R. "Interplay of $\pi$-Electron Delocalization and Strain in [n](2,7)Pyrenophanes." *Journal of Organic Chemistry* 2008, *73*, 8001–8009.

Doering, W. v. E., and Zeiss, H. H. "Methanolysis of Optically Active Hydrogen 2,4-Dimethylhexyl-4-phthalate." *Journal of the American Chemical Society* 1953, *75*, 4733–4738.

Einstein, A. "On the Method of Theoretical Physics." *Philosophy of Science* 1934, *1*, 163–169.

Fukui, K. *Theory of Orientation and Stereoselection*. Berlin: Springer-Verlag, 1975.

Goering, H. L., Briody, R. G., and Levy, J. F. "The Stereochemistry of Ion Pair Return Associated with Solvolysis of p-Chlorobenzhydryl p-Nitrobenzoate." *Journal of the American Chemical Society* 1963, *85*, 3059–3061.

Goering, H. L, and Hopf, H. "Stereochemistry of Ion-Pair Return Associated with Solvolysis of para-Substituted Benzyhydryl p-nitrobenzoates." *Journal of the American Chemical Society* 1971, *93*, 1224–1230.

Gross, D. J. "The Role of Symmetry in Fundamental Physics." *Proceedings of the National Academy of Sciences USA* 1996, *93*, 14256–14259.

Hammett, L. P. *Physical Organic Chemistry: Reaction Rates, Equilibria, and Mechanisms*. New York: McGraw-Hill, 1940.

Hammond, G. S. "Physical Organic Chemistry after 50 Years: It Has Changed, But Is It Still There?" *Pure and Applied Chemistry* 1997, *69*, 1919–1922.

Hare, S. R., and Tantillo, D. J. "Post-transition State Bifurcations Gain Momentum—Current State of the Field." *Pure and Applied Chemistry* 2017, *89*, 679–698.

Hartmann, H., and Longuet-Higgens, H. C. "Erich Hückel. 9 August 1896–16 February 1980." *Biographical Memoirs of Fellows of the Royal Society* 1982, *28*, 152–162.

Heisenberg, W. *Physics and Philosophy: The Revolution in Modern Science*. London: Penguin, 2000.

Hofstadter, D. R. *Gödel, Escher, Bach: An Eternal Golden Braid*. New York: Basic Books, 1999.

Houk, K. N., Gonzalez, J., and Li, Y. "Pericyclic Transition States: Passion and Punctilios, 1935–1995." *Accounts of Chemical Research* 1995, *28*, 81–90.

Houk, K. N., and Liu, P. "Using Computational Chemistry to Understand and Discover Chemical Reactions." *Daedalus* 2014, *143*, 49–66.

Houk, K. N., Liu, F., Yang, Z., and Seeman, J. I. "Evolution of the Diels-Alder Reaction Mechanism Since the 1930s: Woodward, Houk with Woodward, and the Influence of Computational Chemistry on Understanding Cycloadditions." *Angewandte Chemie International Edition* 2021, *60*, 12660–12681.

Judson, H. F. *The Eighth Day of Creation: Makers of the Revolution in Biology*. New York: Simon and Shuster, 1979.

Kermack, W. O., and Robinson, R. L. "An Explanation of the Property of Induced Polarity of Atoms and an Interpretation of the Theory of Partial Valencies on an Electronic Basis." *Journal of the Chemical Society, Transactions* 1922, *121*, 427–440.

Kikuchi, S. "A History of the Structural Theory of Benzene—The Aromatic Sextet Rule and Hückel's Rule." *Journal of Chemical Education* 1997, *74*, 194–201.

Klein, D. *Organic Chemistry*. Hoboken, NJ: John Wiley, 2012.

Kroto, H. W., Heath, J. R., O'Brien, S. C., Curl, R. F., and Smalley, R. E. "$C_{60}$: Buckminsterfullerene." *Nature* 1985, *318*, 162.

Kroto, H. "$C_{60}$, Fullerenes, Giant Fullerenes, and Soot." *Pure and Applied Chemiestry.* 1990, *62*, 407–415.

Krygowski, T. M., and Cyranski, M. K. "Structural Aspects of Aromaticity." *Chemical Reviews* 2001, *101*, 1385–1419.

Krygowski, T. M., and Szatylowicz, H. "Aromaticity: What Does It Mean?" *ChemTexts* 2015, *1*, 12.

Kuhn, T. S. *Black-Body Theory and the Quantum Discontinuity, 1894–1912*. Chicago: University of Chicago Press, 1987.

Kuhn, T. S. *The Structure of Scientific Revolutions*. 4th ed.; Chicago: University of Chicago Press, 2012.

Leffek, K. T. *Sir Christopher Ingold: A Major Prophet of Organic Chemistry*. Nova Lion: Victoria, BC, Canada, 1996.

Lenoir, D., and Tidwell, T. T. "The History and Triumph of Physical Organic Chemistry." *Journal of Physical Organic Chemistry* 2018, *31*, e3838.

Lowry, T. H., and Richardson, K. S. *Mechanism and Theory in Organic Chemistry*. 3rd ed.; New York: Harper and Row, 1987.

Mayr, H. "Physical Organic Chemistry—Development and Perspectives." *Israel Journal of Chemistry* 2016, *56*, 30–37.

Miller, B. *Advanced Organic Chemistry: Reactions and Mechanisms*. 2nd ed. Upper Saddle River, NJ: Pearson Prentice Hall, 2004.

Nestoros, E., and Stuparu, M. C. "Corannulene: A Molecular Bowl of Carbon with Multifaceted Properties and Diverse Applications." *Chemical Communications* 2018, *54*, 6503–6519.

Nordlander, J. E., Owuor, P. O., and Haky, J. E. "Regiochemistry of the Addition of DCI to trans-1,3-Pentadiene." *Journal of the American Chemical Society* 1979, *101*, 1288–1289.

Paradisi, C., and Bunnett, J. F. "Strong Dependence of the Incidence of Internal Return during Solvolysis of Sec-alkyl Benzenesulfonates on the Structure of the Alkyl Group." *Journal of the American Chemical Society* 1981, *103*, 946–948.

Pauling, L. *The Nature of the Chemical Bond and the Structure of Molecules and Crystals: An Introduction to Modern Structural Chemistry*. 3rd ed. Ithaca, NY: Cornell University Press, 1960.

Popper, K. R. *The Logic of Scientific Discovery*. London: Hutchinson, 1959.

Raber, D. J., Bingham, R. C., Harris, J. M., Fry, J. L., and Schleyer, P. V. R. "Role of Solvent in the Solvolysis of tert-Alkyl Halides." *Journal of the American Chemical Society* 1970, *92*, 5977–5981.

Roberts, J. D. "The Beginnings of Physical Organic Chemistry in the United States." *Bulletin of the History of Chemistry* 1996, *19*, 48–56.

Saunders, W. H. J., and Edison, D. H. "Mechanisms of Elimination Reactions. IV. Deuterium Isotope Effects in E2 Reactions of Some 2-Phenylethyl Derivatives." *Journal of the American Chemical Society* 1960, *82*, 138–142.

Schreiner, P. R. "Quantum Mechanical Tunneling Is Essential to Understanding Chemical Reactivity." *Trends in Chemistry* 2020, *2*, 980–989.

Schreiner, P. R., Reisenauer, H. P., Ley, D., Gerbig, D., Wu, C.-H., and Allen, W. D. "Methylhydroxycarbene: Tunneling Control of a Chemical Reaction." *Science* 2011, *332*, 1300–1303.

Schreiner, P. R., Reisenauer, H. P., Pickard Iv, F. C., Simmonett, A. C., Allen, W. D., Matyus, E., and Csaszar, A. G. "Capture of Hydroxymethylene and Its Fast Disappearance through Tunneling." *Nature* 2008, *453*, 906–909.

Schultz, G. "Feier der Deutschen Chemischen Gesellschaft zu Ehren August Kekulé's." *Berichte der Deutschen Chemischen Gesellschaft* 1890, *23*, 1265–1312.

Seeman, J. I. "Woodward–Hoffmann's Stereochemistry of Electrocyclic Reactions: From Day 1 to the JACS Receipt Date (May 5, 1964 to November 30, 1964)," *Journal of Organic Chemistry* 2015, *80*, 11632–11671.

Siegel, J. S. "Mills-Nixon Effect: Wherefore Art Thou?" *Angewandte Chemie International Edition* 1994, *33*, 1721–1723.

Singham, M. "The Copernican Myths" *Physics Today* 2007, *60*, 48–52.

Streitwieser, A. *A Lifetime of Synergy with Theory and Experiment*. Washington, DC: American Chemical Society, 1997.

Streitwieser, A. "Stereochemistry of the Primary Carbon. II. Esters of Optically Active Butanol-1-d. Solvolysis of Butyl-1-d Brosylate." *Journal of the American Chemical Society* 1955, *77*, 1117–1122.

Streitwieser, A., Heathcock, C. H., and Kosower, E. M. *Introduction to Organic Chemistry*. 4th ed. New York: Macmillan, 1992.

Streitwieser, A., and Schaeffer, W. D. "Stereochemistry of the Primary Carbon. III. Optically Active 1-Aminobutane-1-d1,2." *Journal of the American Chemical Society* 1956, *78*, 5597–5599.

Streitwieser, A., and Schaeffer, W. D. "Stereochemistry of the Primary Carbon. IV. The Decomposition of Optically Active 1-Butyl-1-d Chlorosulfite." *Journal of the American Chemical Society* 1957, *79*, 379–381.

Tantillo, D. J. "Dynamic Effects on Organic Reactivity—Pathways to (and from) Discomfort." *Journal of Physical Organic Chemistry* 2021, *34*, e4202.

Tenud, L., Farooq, S., Seibl, J., and Eschenmoser, A. "Endocyclische SN-Reaktionen am gesättigten Kohlenstoff? Vorläufige Mitteilung." *Helvetica Chimica Acta* 1970, *53*, 2059–2069.

Thomas, J. B., Waas, J. R., Harmata, M., and Singleton, D. A. "Control Elements in Dynamically Determined Selectivity on a Bifurcating Surface." *Journal of the American Chemical Society* 2008, *130*, 14544–14555.

Vogel, P., and Houk, K. N. *Organic Chemistry: Theory, Reactivity and Mechanisms in Modern Synthesis*. Weisheim, Germany: Wiley-VCH, 2019.

Wang, Z., Hirschi, J. S., and Singleton, D. A. "Recrossing and Dynamic Matching Effects on Selectivity in a Diels-Alder Reaction." *Angewandte Chemie International Edition* 2009, *48*, 9156–9159.

Watson, J. D., and Crick, F. H. "Molecular Structure of Nucleic Acids: a Structure for Deoxyribose Nucleic Acid." *Nature* 1953, *171*, 737–738.

Weininger, S. J. "Benzene and Beyond: Pursuing the Core of Aromaticity." *Annals of Science* 2015, *72*, 242–257.

Winstein, S., Gall, J. S., Hojo, M., and Smith, S. "Racemization, Acetolysis and Radiochloride Exchange of Two Alky; Chloride." *Journal of the American Chemical Society* 1960, *82*, 1010–1011.

Woodward, R. B., and Hoffmann, R. "The Conservation of Orbital Symmetry." *Angewandte Chemie International Edition* 1969, *8*, 781–853.

Zeki, S., Romaya, J., Benincasa, D., and Atiyah, M. "The Experience of Mathematical Beauty and Its Neural Correlate." *Frontiers in Human Neuroscience* 2014, *8*, 68.

Zimmerman, H. "Möbius-Hückel Concept in Organic Chemistry. Application of Organic Molecules and Reactions." *Accounts of Chemical Research* 1971, *4*, 272–280.

# Index